Studies in Computational Intelligence

Volume 818

The series "Studies in Computational Intelligence" (SCI) publishes new developments and advances in the various areas of computational intelligence—quickly and with a high quality. The intent is to cover the theory, applications, and design methods of computational intelligence, as embedded in the fields of engineering, computer science, physics and life sciences, as well as the methodologies behind them. The series contains monographs, lecture notes and edited volumes in computational intelligence spanning the areas of neural networks, connectionist systems, genetic algorithms, evolutionary computation, artificial intelligence, cellular automata, self-organizing systems, soft computing, fuzzy systems, and hybrid intelligent systems. Of particular value to both the contributors and the readership are the short publication timeframe and the world-wide distribution, which enable both wide and rapid dissemination of research output.

The books of this series are submitted to indexing to Web of Science, EI-Compendex, DBLP, SCOPUS, Google Scholar and Springerlink.

More information about this series at http://www.springer.com/series/7092

Jan W. Owsiński

Data Analysis in Bi-partial Perspective: Clustering and Beyond

Springer

Jan W. Owsiński
Systems Research Institute
Polish Academy of Sciences
Warsaw, Poland

ISSN 1860-949X ISSN 1860-9503 (electronic)
Studies in Computational Intelligence
ISBN 978-3-030-13391-7 ISBN 978-3-030-13389-4 (eBook)
https://doi.org/10.1007/978-3-030-13389-4

Library of Congress Control Number: 2019931851

This Springer imprint is published by the registered company Springer Nature Switzerland AG
The registered company address is: Gewerbestrasse 11, 6330 Cham, Switzerland

Foreword

The considerations forwarded in the present volume suggest that the use of the general "bi-partial" objective function leads to a more effective capacity of solving the problems of cluster analysis in that it implies obtaining both the cluster content and the cluster number, without recurring to an "external" criterion. The method, proposed, embraces also the possibility of designing a relatively simple and intuitive algorithm of sub-optimisation. The same principles can be similarly successfully applied to several other essential problems of data analysis.

As a "by-product" of the developments, related to the paradigm of the bi-partial objective function, several important issues surface in a natural manner and get an interesting interpretation and/or explanation. One of those is the matter of "scale", crucial for the understanding of the adequacy of partitions in clustering, here intimately associated with the transformation between distance and proximity. Within the bi-partial approach, this "scale" can be made explicit and subject to choice.

Another such "by-product" is constituted by the association with the progressive merger (agglomerative) techniques, classical to clustering, the respective merger rules, minimum distance principle, etc. Here, definitely, the bi-partial approach provides a much deeper insight into the principles and structures involved.

Regarding clustering, the book considers solely the clustering problem proper, so that a lot of related research issues are left untreated, like, the problem of mixture of distributions or the specific problems associated with pattern recognition. On the other hand, the content of this book does not address at all the currently dominating meta-heuristic approaches, applied in clustering, since they refer uniquely to the algorithmic side of finding the solutions to the clustering problem, and all of them have, anyway, to be guided by the more general principles as to what such solutions actually are. And this is exactly what the study here reported deals with.

It should be noted that the book is "technical" in the sense that it occupies a place between computer science type of considerations and those belonging to applied mathematics, emphasis being on practical and effective solving of problems having definite substance matter contents, whose interpretation is of primary importance.

An important reservation ought to be made at this point: ***the book is not about yet another objective function for clustering***. It is about an integral approach, in which the proposed form of the general objective function, which can be instantiated for the various concrete perceptions and interpretations of the clustering problem, plays the central role.

Few words are perhaps due, concerning the relation between the content of this book and the so-called soft approaches. The adjective "soft" appears, namely, to be reserved to a definite set of methodologies, so that, for example, in clustering, the classical k-means algorithm is treated as "hard", when compared to fuzzy c-means (FCM), qualified as "soft". This is, of course, a false image, since clustering as such should be in many instances considered to be a "soft" approach (unless it is not only treated, but actually applied as just a preliminary stage) to such problems as model identification, facility location, cell formation in flexible manufacturing, information retrieval and many others. For virtually all of these, mathematical programming formulations exist or can relatively easily be developed, possibly accompanied by sensitivity analysis. Hence, a difference between these two ways of approaching the respective problems can easily be established. Even within the data analysis domain, most of clustering approaches (including naturally those based on fuzzy or intuitionistic precepts) should be considered "soft" in the sense that the respective problem and the way to solve it are not very precisely defined, with a wide margin for interpretation of both problem statement and the results obtained.

The content of this report has the following structure: first, notations used in the study are introduced and their context ("problem background") is explained in Chap. 1. Then, in Chap. 2, the generic problem of cluster analysis is formulated, along with all the consequences thereof, especially the indications of necessary complements, needed to make the problem of clustering practically tractable.

In Chap. 3, the general formulation of the objective function is introduced and justified, along with some basic illustrations and examples. The rationale for and the ways of constructing this objective function are amply illustrated with examples of its concrete implementations, satisfying the prerequisites previously proposed, in Chap. 4 for a number of diverse standard problems in data analysis, but also bordering upon other domains, e.g. operational research. These examples are then complemented in Chap. 5 by some further instances, belonging, however, uniquely to the domain of cluster analysis. This chapter also contains the consideration of other criteria and evaluation functions, used in various problems of data analysis, which are related to the here introduced general two-sided objective function and to derivation of the connections between them. The chapter ends with some remarks and suggestions on the way the clustering methods and results ought to be assessed.

The algorithm, which sub-optimises with respect to the general objective function, is introduced, against the background of the required properties, along with the concrete versions of this algorithm for a spectrum of the exemplary implementations of the objective function, in Chap. 6. It is also shown there that the algorithms obtained for the bi-partial objective function share some of the essential characteristics with the classical hierarchical progressive merger algorithms, described with the well-known Lance–Williams formula. Being in a way similar to those algorithms, the ones proper for the bi-partial objective function provide for the stop condition, concerning the merger iterations, as well as a natural index of hierarchy.

In Chap. 7, the application is shown of the entire approach and its philosophy, including also the possibility of deriving an effective algorithm, to quite a specific, and partly different case, namely the one of preference (precedence) aggregation. Following the final remarks, forming Chap. 8, and a concise index, the volume ends with a broad bibliography, related, on the one hand, primarily to cluster analysis and the works using or based on explicit formulations of objective functions, and on the other hand—to those other issues in data analysis, where similar principles are observed or at least postulated. A separate list of references is put together, concerning the work of the present author, essentially concerning the bi-partial approach, so as to not make an impression of dominating the references altogether.

The book is meant for data science specialists, who would like to broaden their perspective on the fundamental approaches available, but first of all—on the way many problems are perceived in and thus to find answers to some questions usually either overlooked or solved via cumbersome or not fully convincing manners. It is also meant for graduate students, dealing with computer and data sciences, who might wish to complement their knowledge and skills with a fresh insight into many problems, otherwise treated in some standard "academic" manners.

The Readers, who might be interested in the "historical" roots of the approach here presented, are encouraged to take, perhaps, first a look at the Sect. 5.4.3 of the volume, especially if they are already somewhat knowledgeable in the domain of clustering.

Warsaw, Poland Jan W. Owsiński

Preface

The present book results from the studies, conducted by the author, concerning data analysis, and especially cluster analysis and preference aggregation. The volume contains the general formulation, the properties, the examples and the techniques associated with a general objective function, referred to as "bi-partial" objective function, devised mainly for purposes of effective solving of the clustering problems.

This objective function is based on the principle of simultaneous consideration of two aspects of clustering: the intra-cluster similarity and inter-cluster dissimilarity (or, dually, intra-cluster distance and inter-cluster proximity). This sounds, definitely, very much obviously, and even perhaps bordering upon triviality. Yet, it is shown here how the general principles of construction of such an objective function can be implemented through concrete, practically applicable formulations. Further, it is demonstrated that both the general form of the function and its concrete implementations imply the solutions to the clustering problem in terms of both the number of clusters and their composition. Then, a general algorithm is proposed, leading to the sub-optimal solutions with respect to the objective function, through a special type of classical progressive merger procedure. The properties of this algorithm and the examples of its concrete implementations are also presented.

The key property that is stressed in the methodology here proposed is the adequate representation of the generic problem of cluster analysis. It is shown that most, if not all of the existing approaches fail with this respect, and that is why they usually also fail to propose, within the same approach, a consistent and effective algorithm of finding the solutions to the clustering problem.

It is also shown how the general objective function can be formulated in a relatively easy, but also constructive and effective manner for quite a wide variety of problems in multivariate data analysis (e.g. categorisation, optimum histogram and rule extraction). Likewise, examples are shown of its equivalence to some kinds of quality criteria or indices used also, or at least referred to in various kinds of data analytic problems. In particular, application of the fundamental principles of construction of the objective function and of the algorithm to the problem of preference aggregation is presented.

Warsaw, Poland Jan W. Owsiński

Contents

List of Figures

List of Tables

Chapter 1
Notation and Main Assumptions

1.1 Notation

First, we will introduce the list of main notations used in the report. The sequence, in which the notations are here listed, follows roughly the sequence, in which they appear in the text:

i	Index of an object (observation), j is also used as object index,
n	Total number of objects indexed i,
I	Set of object indices, $I = \{1,\ldots,i,\ldots,n\}$, i.e. $i \in I$,
k	Index of a variable (attribute, feature), l is also used as variable index,
m	Total number of variables considered,
K	Set of variable indices, $K = \{1,\ldots,k,\ldots,m\}$, i.e. $k \in K$,
x_{ik}	Value, taken on by variable k for object i,[1]
X	Matrix of object by variable data, $X = \{x_{ik}\}_{ik}$, where objects correspond to rows and variables correspond to columns,
x_i	Description of an object in terms of variable values, $x_i = \{x_{ik}\}_k = \{x_{i1},\ldots,x_{ik},\ldots,x_{im}\}$,
X_I	Object descriptions, $X_I = \{x_i\}$, when treated as equivalent to the description of a set (I) of objects,
A, B	General designations, possibly accompanied by the appropriate subscripts or superscripts, of the sets of objects, *Note that whenever objects considered belong to the set X_I, they are usually identified not by their complete descriptions x_i, but, simply, by their indices, $i \in I$, and so also the sets of objects,*

[1]In this book we do not deal at all with the issue of missing observations (missing values of x_{ik}), just as we do not deal with the potential various kinds of uncertainties, associated with these values.

J. W. Owsiński, *Data Analysis in Bi-partial Perspective: Clustering and Beyond*, Studies in Computational Intelligence 818,
https://doi.org/10.1007/978-3-030-13389-4_1

when $\subseteq X_I$, *are equated with the corresponding sets of object indices;*

P — Partition of the set of objects X_I (actually: of the set of object indices I) into subsets, which are denoted A_q, and referred to as "clusters" ; and so, $P = \{A_q\}_q$,

A_q — Subsets ("clusters") of the set of objects, defined in terms of object indices, $A_q \subseteq I$, we shall assume throughout that $\cup_q A_q = I$, meaning that clusters forming any (feasible) partition exhaust the set of objects X_I,

q — Index of a subset ("cluster") of the set of object indices, $q = 1,\ldots,p$, where p is the number of subsets forming a partition P; in case of necessity the number of clusters for a partition P can be denoted $p(P)$,

$x_{.k}$ — Values of the kth variable for all objects, i.e. $x_{.k} = \{x_{ik}\}_i = \{x_{1k},\ldots,x_{ik},\ldots,x_{nk}\}$,

E_K — Space (set) of admissible values of x_i, i.e. of all the potential object descriptions (locations in E_K), so that $x_i \in E_K \; \forall \, i \in l$, and $X_I \subseteq E_K$,

E_k — Space (set, interval) of admissible values of a variable k, so that $x_{ik} \in E_k$, and, except for some special situations and constraining conditions, $E_K = E_1 \times E_2 \times \ldots \times E_k \times \ldots \times E_m$,

x — General designation, possibly accompanied by appropriate subscripts or superscripts, of an object $\in E_K$, but $\notin X_I$, in general, with $x = \{x_{x1},\ldots,x_{xk},\ldots,x_{xm}\}$,

$d(.,.)$ — Distance (dissimilarity) function between a pair of objects: $E_K \times E_K \to R^+ \cup \{0\}$,

$s(.,.)$ — Proximity (similarity) function between a pair of objects: $E_K \times E_K \to R^+ \cup \{0\}$,

$\mathbf{d} = \{d_{ij}\}_{ij}$ — Matrix of distances for objects belonging to X_I,

$\mathbf{s} = \{s_{ij}\}_{ij}$ — Matrix of proximities for objects belonging to X_I,

$d(x_i, x_j)$ — Distance value for objects x_i and x_j, when x_i and $x_j \in X_I$, then a simplified notation is used: $d(i, j)$ or even d_{ij},

$s(x_i, x_j)$ — Proximity (similarity) value for objects x_i and x_j, when x_i and $x_j \in X_I$, then a simplified notation is used: $s(i, j)$ or even s_{ij},

$d^k(.,.)$ — Distance (dissimilarity) function between objects, defined for a single variable $k : E_k \times E_k \to R^+ \cup \{0\}$,

$s^k(.,.)$ — Proximity (similarity) function between objects, defined for a single variable $k : E_k \times E_k \to R^+ \cup \{0\}$,

$d^k(x_i, x_j)$ — Distance value for objects x_i and x_j, defined for a single variable k, when x_i and $x_j \in X_I$, then a simplified notation is used: $d^k(i,j)$ or even d^k_{ij},

$s^k(x_i, x_j)$ — Proximity value for objects x_i and x_j, defined for a single variable k, when x_i and $x_j \in X_I$, then a simplified notation is used: $s^k(i,j)$ or even s^k_{ij},

$D(.,.)$	Distance function between two sets of objects, with values in $R^+ \cup \{0\}$,
$D(.)$	Distance function defined for a single set of objects, with values in $R^+ \cup \{0\}$, for simplicity, notation used is the same as for distance between two sets of objects,
$S(.,.)$ and $S(.)$	Analogous notations for proximities (similarities) between two sets of objects and for individual sets of objects, also with values in $R^+ \cup \{0\}$,
E_P	Space of admissible partitions, that is—the ones, composed of clusters jointly exhausting the set of objects X_I and possibly fulfilling other conditions,
H	Hierarchy (of partitions), defined as a set of partitions, $\{P^t\}_t$, such that $P^t \subset P^{t+1}\ \forall t$, where $P \subset P', P = \{A_q\}_q, q = 1,\ldots,p(P)$, $P' = \left\{A'_{q'}\right\}_{q'}, q' = 1,\ldots,p(P')$, whenever for each A_q there exists such $A'_{q'}$ that $A_q \subseteq A'_{q'}$ and there exists at least one A_q and a corresponding $A'_{q'}$ such that $A_q \subseteq A'_{q'}$, meaning that $\mathrm{card}A_q < \mathrm{card}A'_{q'}$; where card A denotes cardinality of the set A,
E_H	Space of *admissible* hierarchies, that is—of hierarchies composed of admissible partitions,
$C(P)$, $Q(P)$	Criteria or objective functions defined over the space of partitions;
$C_H(H)$, $Q_H(H)$	Analogous criteria or objective for hierarchies.

1.2 Situation Studied and Its Characterisation

We consider a set of distinct objects, or observations, indexed i, whose number is n, and the set of object indices is denoted I, so $I = \{1,\ldots,i,\ldots,n\}$. These objects are described by definite variables (characteristics, features), which are numbered with index k, and we consider altogether m such variables. The set of variable indices is denoted K, $K = \{1,\ldots,k,\ldots,m\}$.

Particular variables, indexed k, take on values from the sets denoted E_k. It should be emphasised that we do not assume anything about the "variable spaces" E_k, except for the fact that their elements are constituted by values, on which we should be able to perform only some basic arithmetic and/or logic operations, to which we shall yet return.

Thus, object descriptions, denoted x_i, are composed of variable values, denoted x_{ik}, $x_{ik} \in E_k$, so that $x_i = \{x_{i1},\ldots,x_{ik},\ldots,x_{im}\}$. We can further say that $x_i \in E_K$, where E_K is the space, or the set, of all the possible object descriptions. For virtually all practicable cases we can, and indeed we do, assume that $E_K = E_1 \times E_2 \times \ldots \times E_k \times \ldots \times E_m$.

The values of x_{ik} form an $n \times m$ matrix X. Rows of this matrix, indexed i, correspond to object descriptions, x_i, while its columns—to the values of particular variables k, taken by the objects considered. We shall denote the (values in the) columns by $x_{.k}$, with $x_{.k} = \{x_{1k}, \ldots, x_{ik}, \ldots, x_{nk}\}$.

When speaking of the object descriptions as a set of x_i, rather than a matrix, we shall be using notation X_I, so that $x_i \in X_I \subseteq E_K$. So, we deal with the data given as X or X_I.

The space of object descriptions, E_K, contains, of course, at least potentially, the objects not belonging to the set X_I. When considering such objects not from X_I, or, in general, any objects in E_K, we will be using notations like x, x', etc.

Yet, we shall be primarily concerned with the set X_I, being the set of given data items. So, it will be possible to use in many cases the indices i as uniquely defining the objects and their sets. The set X_I shall therefore be identified with the set I.

Of our primary concern will be with the partitions of the set X_I of objects, that is, equivalently—of the set of their indices I. A partition P of I is a set of subsets $A_q \subseteq I, P = \{A_q\}_q$, these subsets being indexed $q = 1, \ldots, p$, where p is the cardinality of partition, such that $\cup_q A_q = I$, i.e. the partitions here considered are exhaustive, or cover the set I. If not otherwise stated, it will be assumed that $A_q \subseteq P$ are all nonempty. If an emphasis on dependence of p on partition is necessary, we shall be using for the number of subsets in a partition P the notation $p(P)$.

It is the subsets A_q, forming a partition, that are referred to as "clusters", such an expression being justified by the assumed character of the subsets and partitions sought in the "clustering problem", as this problem is formulated here later on in Chap. 2.

Throughout this volume we shall also mainly be concerned with the proper partitions, that is, the ones, for which, in addition to the condition of exhausting the set I, the condition $A_q \cap A_{q'} = \emptyset \ \forall q, q'$ holds, that is—the clusters, forming a proper partition, are all mutually disjoint. In fact, if not otherwise stated, the partitions considered in the book shall be proper, although the more general case of overlapping and fuzzy clusters shall also be considered (treatment of fuzzy clusters necessitates introducing additional notions, and so, to keep the notation possibly simple, we shall introduce them at a proper place in the book).

Whenever other assumptions are made on the partitions considered, of which there are few that are indeed quite important, they are explicitly stated.

Partitions may form hierarchies, denoted H. A set of partitions, $\{P^t\}_t$, forms a hierarchy, when condition $P^t \subset P^{t+1} \forall t$ is satisfied. We say that $P \subset P'$, when for each cluster A_q, forming partition P, there exists a cluster $A'_{q'}$ in P' such that $A_q \subseteq A'_{q'}$, and for at least one A_q there is $A_q \subset A'_{q'}$, the latter meaning that $\text{card}A_q < \text{card}A'_{q'}$.

The space of hierarchies, understood in this manner, will be denoted E_H.

There exists a special kind of hierarchies, in which $P^1 = I$, that is: $A^1_q = q$, $q = 1, \ldots, p = n$, i.e. each object constitutes a separate cluster in the initial (t = 1) partition. In this kind of hierarchies we have $t = 1, \ldots, T$, and $P^T = \{I\}$, that is:

$p(P^T) = 1, P^T = A_1^T = \{1,\ldots,n\}$. This means that the final partition in the hierarchy is composed of just one cluster, containing all objects from I. We shall deal to a large extent with this kind of hierarchies in the considerations here presented.

Now, between any two objects from E_K, whether belonging to X_I, or not, two functions may, and here actually will be defined, the function of distance (dissimilarity), and the function of proximity (similarity, affinity). These two functions are defined as $d(.,.) : E_K \times E_K \rightarrow R^1 \cup \{0\}$ and $s(.,.) : E_K \times E_K \rightarrow R^1 \cup \{0\}$, respectively. Their values for pairs of objects belonging to X_i will be denoted $d(x_i, x_j)$, or $d(i, j)$, or simply d_{ij}, and $s(x_i, x_j)$, or $s(i, j)$, or simply s_{ij}, respectively.

We shall speak of distance and proximity when the functions involved satisfy, in addition to the above, the following conditions:

$$\begin{array}{ll} d(x,x') = d(x',x)\ \forall\ x,x' \in E_K & \text{(symmetry)} \\ d(x,x) = 0\ \forall\ x \in E_K & \text{(identity)} \end{array}$$

and

$$\begin{array}{ll} s(x,x') = s(x',x)\ \forall x,x' \in E_K & \text{(symmetry)} \\ s(x,x) = \max s(x',x'';x',x'' \in E_K)\ \forall x \in E_K & \text{(identity).} \end{array}$$

No other conditions, in particular those related to the otherwise often imposed metricity, will be imposed on these two functions.

In case, however (which, indeed, **is** the case of this study), when both distances and proximities are defined simultaneously for the same set of objects, we impose a mild condition

$$d(x,x') \le d(x'',x''') \Leftrightarrow s(x,x') \ge s(x'',x''') \quad \forall x,x',x'',x''' \in E_K.$$

It is assumed that distance and proximity between objects are either based on analogous notions, defined for individual variables, or are closely associated with such notions, i.e.

$$d^k(.,.) : E_k \times E_k \rightarrow R^1 \cup \{0\}$$

and

$$s^k(.,.) : E_k \times E_k \rightarrow R^1 \cup \{0\},$$

respectively, these notions being referred to as partial distance and partial proximity. The values of these two functions for objects belonging to X_I are denoted, analogously as before, $d^k(x_i,x_j)$, $d^k(i,j)$ or d^k_{ij}, and $s^k(x_i,x_j)$, $s^k(i,j)$, or s^k_{ij}, respectively, and these values satisfy the same conditions as set on distance and proximity. For objects, which may not belong to the set X_I, the general notation, like $d(x, x')$ and $s(x, x')$, and also $d^k(x, x')$ and $s^k(x, x')$, will be used.

The way, in which partial distance and proximities are aggregated to form proper inter-object distance and proximities, is accordingly defined for each case, and is of little interest here.

Distances and proximities calculated for objects belonging to X_I form, therefore, conform to the conditions assumed, symmetric matrices, denoted $\boldsymbol{d} = \{d_{ij}\}_{ij}$ and $\boldsymbol{s} = \{s_{ij}\}_{ij}$.

Let us only note at this point that we have assumed very little indeed about the E_k and E_K, and so doubts may arise as to the capacity of practical definitions of distance and proximities for these spaces. Yet, in view of the—simultaneously—very restricted requirements with respect to the definitions of distance and proximity, it can safely be admitted that for virtually all kinds of E_k, including the nominal ones, and so also for virtually all kinds of E_K, as their aggregates, both distances and proximities can be sensibly defined.[2]

For purposes of this study we introduce a general notation for distances and proximities between sets of objects, $D(.,.)$ and $S(.,.)$, respectively. We shall use, however, the same notation for the distances and proximities characterising single sets of objects, $D(.)$ and $S(.)$. (An example of such distance, characterising a set of objects, is provided by its diameter, i.e. maximum distance between any two objects in the set.) Of these distances and proximities we shall for a while only assume that their values are nonnegative, but also, as previously, in case both are simultaneously defined, which **is** the case here, we require

$$D(A,A') \leq D(B,B') \Leftrightarrow S(A,A') \geq S(B,B')$$

for all sets A, A', B, B' of objects $\in E_K$.

A similar requirement, though, is not, at this point, imposed on the distance and proximity functions relative to single sets, i.e. $D(.)$ and $S(.)$ (there is, definitely, no strict analogy to the case of two entities).

As of now, we will not introduce any more precise assumptions about D and S, except for a general statement that they satisfy a reasonable monotonicity with regard to, respectively, the distances and proximities within the argument sets. The more detailed assumptions will be specified as need arises. It will, however, be assumed throughout the report that S and D, whenever defined and used, appropriately reflect the respective proximities and distances within or between the sets of objects, conform to the principles of such representation, explicitly adopted for a given approach, using S and/or D.

We will also use notation for criteria or objective functions, related to entire partitions, namely $C(P)$ and $Q(P)$. The distinction between the two will be explained later on. Analogous functions can be defined for hierarchies, $C_H(P)$ and $Q_H(P)$.

[2]We shall not deal here at all with the possible definitions of distances and/or proximities for various kinds of variables, as there exists a vast literature on this subject, and the characteristics of the particular distance definitions have little or, indeed, no significance for the considerations here forwarded.

All of the above notations will be, as necessity arises, complemented with appropriate subscripts or superscripts, explained each time they would appear.

Finally, let us emphasise that we do not deal here with the issue of preliminary operations on data, and so, in particular, we do not treat nor specify the normalisation procedures, assuming that the data processed are simply given, whether normalised or not. If any difference in respective reasoning may arise, resulting from possible application of such preliminary procedures, of importance for our considerations, this fact will be duly announced and accounted for.

Chapter 2
The Problem of Cluster Analysis

2.1 The General Formulation

We shall now formulate in general terms the basic or "generic" problem of cluster analysis, and then discuss the consequences of this formulation. Thus:

> The problem of clustering is defined as follows:
>
> given the set of objects, indexed i, $i \in I = \{1, \ldots, n\}$, characterised by variables indexed k, $k \in k = \{1, \ldots, m\}$, the object descriptions being given by $x_i = \{x_{i1}, \ldots, x_{ik}, \ldots, x_{im}\}$, composed of the values for particular variables, partition the set of objects into subsets (equivalent to subsets of their indices) A_q, $q = 1, \ldots, p$, such that $\cup_q A_q = I$, so that in the same subsets A_q possibly similar (close) objects are contained, while in different subsets A_q—possibly distant (dissimilar) ones.

This basic and "generic" formulation is intuitively obvious and appealing. It is also generally accepted in the literature. In the study here presented we shall stress the need of adequately representing the above formulation in its entirety and the resulting capacity of effective solving the clustering problem.

Yet, it is also obvious that this general formulation is grossly insufficient in terms of precision to allow for its practical treatment. In what follows we shall shortly discuss the issues that arise when one tries to pass from this general formulation to the one that could be practically solved, or to the solution proper. Some of these issues have been touched upon in the introductory chapter, but also for them we shall give here appropriate comments.

At the beginning, though, we shall forward some general comments on the meaning of the problem and its solution and the broader consequences thereof.

J. W. Owsiński, *Data Analysis in Bi-partial Perspective: Clustering and Beyond*, Studies in Computational Intelligence 818,
https://doi.org/10.1007/978-3-030-13389-4_2

2.2 The Meaning of the Problem

The problem of clustering is, in fact, equivalent to one of the basic functions of human mind and culture: grouping (and appropriately naming) of the similar and distinguishing the dissimilar. This, actually, is the basis of human language, with all the distinct words and expressions, addressing different groups of objects (whether designated with nouns, adjectives, verbs, etc.).

Figure 2.1 shows a simple illustration for such a language-oriented interpretation of the clustering problem.

Moreover, this problem is also—in a more formalised manner—the basis of the fundamental scientific procedure, even if primarily at its initial stages. Here we can speak of analogies, as well of similarity or identity of processes, models and theories, corresponding to clusters (these clusters containing objects, being individual observations or facts, described according to an adopted, e.g. biological, methodology), and a model or theory, encompassing the entire partition. In the latter case it is the way, in which the partition is done, that would constitute the core of the broader model or theory.

In most usual practice, cluster analysis is an element of the exploratory data analysis, that is—it is applied on the very first stages of a (potential) deeper data analysis procedure, when little, or close to nothing, is known about the data analysed (here, the descriptions x_i, forming the set X). The problem of clustering, as formulated here, namely, contains barely any presumption concerning the character of the data or the process, from which they may have resulted.

By formulating and solving the problem of clustering we try to identify some structure in the data set, this structure having the form of partition into subsets (or a form related to partition, e.g. a hierarchy of partitions). Such a structure is perhaps the most primitive of all the structures that can be imposed on or identified in a data set.

Note, however, that the formulation of the clustering problem we are using here does not refer to an attempt to recognise or uncover any kind of "objective" knowledge about the data set and its structure, like the "models", representing the

Fig. 2.1 An example: furniture objects described by two variables

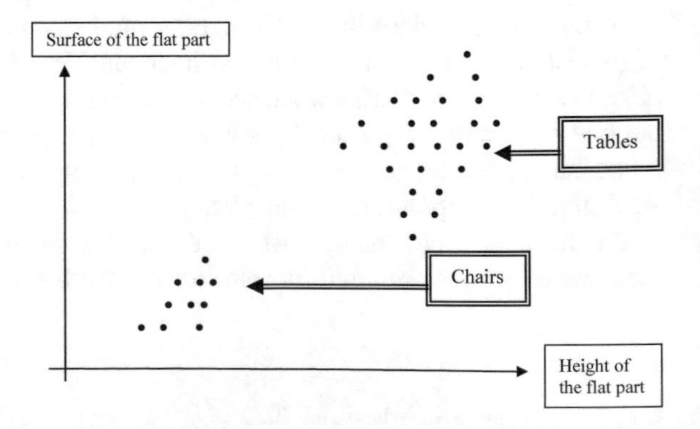

processes, having led to the appearance of data. To the contrary, *the formulation here considered of the clustering problem suggests that a partition might be prescribed for a set of objects even if there were no "objective" division into subsets*. Whether such a situation arises, or we in fact go for a structure that approximates some "objective" one, definitely, depends upon the details of concrete formulation of the problem, going much father into the details than the general formulation we consider, and thus also involving additional, constraining assumptions.

The sole "objectivity" that is implied by the formulation of the clustering problem being here—and certainly also elsewhere—solved is the one, related to the possibly "objectively" good choice of a partition exactly conform to the problem formulation.

Notwithstanding the above very important reservation, in view of the sweeping generality of the initial problem formulation, before we could pass over to the attempt at practical solving of the problem and identification of the structure, we have to specify several elements of the problem more precisely.

2.3 The Issues of Practicality

This rather ample section of the report is devoted to the presentation and discussion of these elements of the clustering problem, which have to be specified in order to be able to practically analyse and solve any concrete task, involving clustering. We shall, however, refrain from too meticulous considerations, focussing on presentation of the essential aspects, which are of importance for the reasoning, forwarded in the volume.

2.3.1 The Nature of Partition

First is the issue of the ultimate structure sought, i.e. *the nature of partition*. As mentioned already, we shall be mainly considering the proper partitions, i.e. the ones, which do not only exhaust the set of objects (object indices), but also the clusters forming the partition are mutually disjoint.

This is, of course, not the only possible form of potential partition sought. The obvious alternative is the one, in which clusters may overlap, i.e. there may exist such q, q' that $A_q \cap A_{q'} \neq \emptyset$. Sometimes in such formulation the subsets A_q are called "cliques" rather than clusters.

Figure 2.2 shows a case, in which formation of overlapping clusters may be fully justified or even advisable.

Then, however, there is a whole, indeed very broad domain of cluster analysis, in which clusters are defined as *fuzzy sets*. Thus, subsets A_q are defined through the set of p membership functions $\mu_q(.), I \rightarrow [0, 1]$, that is, $A*_q = \{\mu_q(x_i)\}_i$, or, more in

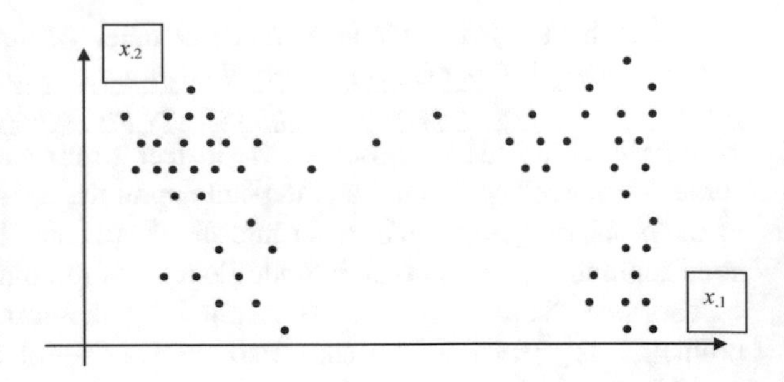

Fig. 2.2 An example of a set of objects, for which overlapping clusters may be justified

accordance with our notation, $\mu_q(i)$, where $A*_q$ denotes a cluster, defined as a fuzzy set, and $\mu_q(i) \in [0, 1]$. A number $\mu_q(i)$ should be interpreted as "degree of membership of the object indexed i in the cluster indexed q, ranging", from 0 (i not belonging to $A*_q$ at all) to 1 (i fully belonging to $A*_q$).

It can easily be seen that in the case like that of Fig. 2.2 the use of fuzzy sets would also be fully justified.

In the above fuzzy set theoretic setting, the case of "usual", the so-called "crisp" clusters, can be considered as a special case, in which, simply, we put the restriction $\mu_q(i) \in \{0, 1\} \ \forall \ i, q$.

Introduction of fuzzy clusters, and thus also of fuzzy partitions, $P*$, entails several issues to be resolved. First, we have already noted that we do not consider empty clusters, meaning, in case of $A*_q$, $\mu_q(i) = 0 \ \forall \ i \in I$. This, however, in the case of fuzzy partitions, is not enough.

First, the condition of exhausting the set of objects, I, takes now the form:

$$\sum_i \sum_q \mu_q(i) = n. \tag{2.1}$$

Another usually imposed condition on fuzzy partitions is as follows:

$$\sum_q \mu_q(i) = 1 \quad \forall i, \tag{2.2}$$

which means that each object should be "distributed" among clusters exactly in its entirety, that is—there is no "part" of an object that is left unassigned to a cluster, and the sum of assignments for an object does not exceed one, for if it did, we would have assigned more than 100% of this object. It is, however, contrary to the latter statement, possible to impose, instead of (2.2),

$$\sum_q \mu_q(i) \geq 1 \ \forall \ i, \tag{2.2a}$$

which is the fuzzy equivalent to admission of "truly overlapping" clusters, or, more appropriately, cliques, in the "crisp" case. Note, though, that fuzzy clusters overlap in a way by their very nature, which is expressed, in particular, through (2.2), with (2.2a) simply extending the possibility of this overlapping up to multiple full membership of an object in several clusters.

While (2.2), or, in some cases, (2.2a), is usually imposed on fuzzy partitions, a condition strengthening the requirement of non-emptiness of $A*_q$ is definitely less frequently strictly applied. When applied, it has most often the form:

$$\sum_i \mu_q(i) \geq 1 \ \forall \ q \tag{2.3}$$

which implies that an admissible fuzzy cluster must altogether represent an equivalent of at least one object. The fact that this condition is not so often applied, results from the potential "conflict" with (2.2), e.g. when an "outlying" object might partly belong to a nearby cluster, and partly form by itself a self-standing cluster, the latter possibility being excluded by (2.3). Thus, if constraint (2.3) is applied, it usually appears in connection with (2.2a).

We shall not go into further considerations, concerning fuzzy partitions, like the possibility of deeper "fuzzyfication" of the cluster descriptions (fuzzy sets of second kind etc.). This results both from the fact that effective clustering algorithms exist for formulations, which largely follow the precepts here outlined, and from the reference being made in these approaches to objective functions of interest to us.

Likewise, we shall not delve into other non-crisp paradigms of data analysis and the respective potential, and, indeed, existing in the literature, techniques of clustering, including the very nature of clusters and of the partition (like, e.g. to what extent the clusters may overlap and what is the respective interpretation, according to, say, the rough sets methodology).

2.3.2 *Similarity Inside and Dissimilarity Outside: How Do the Existing Methods Fare?*

The second, and perhaps most critical aspect, is related to the requirement of putting together the objects that are possibly similar (close) and apart the ones that are possibly dissimilar (distant). Even though this requirement appears to be fully intuitively obvious, its translation into practicable categories is not quite straightforward.

Thus, if we are able of defining the distance function (and/or the proximity function), and the values thereof for the pairs of objects, then, of course, of the objects i, i', i'', for which

$$d(i,i') < d(i',i'') < d(i'',i) \text{ or } s(i,i') > s(i',i'') > s(i'',i)$$

the objects i, i' are more likely to be put together in a cluster than either i', i'' or i, i''. This, however, is far from both any practical assessment as to which partition, as a whole, is better than another one in terms of the clustering problem, and from the establishment of a workable procedure for the determination of a solution.

It is to this issue that the present volume is largely devoted. It proposes a general form of an objective function for the clustering problem, and a practical procedure for sub-optimising this objective function.

Most, if not all, of the existing approaches fail—in a variety of manners—to treat sufficiently explicitly this aspect of the clustering problem, however. Either they provide the basis for comparing partitions within a relatively narrow class, or they concentrate on the practical procedures, without due concern of the possibility of comparing the structures obtained. We shall give short comments on this subject as an illustration, referring to the various types of the existing clustering methods.

Thus, the standard ***progressive merger*** (“***hierarchical clustering***”) procedures use the distance-based reasoning alluded to above in that they proceed as follows:

- **i**. treat all objects, $i \in I$ as n separate clusters ($p = n$),
- **ii**. find in the matrix $\boldsymbol{d}$ the smallest distance among the objects/clusters, d_{ij},[1]
- **iii**. merge the thus found objects/clusters indexed i and j to form one cluster,
- **iv**. is the resulting number of clusters > 1? if yes, go to step v; if no, stop,
- **v**. transform appropriately the distance matrix $\boldsymbol{d}$ in order to accommodate the merger of objects/clusters i and j (it is at this step that the various progressive merger algorithms differ),
- **vi**. go to step ii.

These algorithms do not refer to any explicit objective function, and produce through the above procedure a hierarchy of partitions, in which $P^1 = I$, that is, $p = n$, and $P^n = \{I\}$, i.e. $p = 1$. In order to designate the partition in this hierarchy, which would be treated as a proper solution, some external criterion is used, usually of statistical character. Selection of this external criterion is decisive for the designation of partition, being the solution to the clustering problem.[2]

Another broad group of cluster analytic techniques refers to the generic so-called “***k-means***” procedure. As is known, this procedure is based on the implicit consideration of the objective function, which takes on the general form

$$C(P) = \sum_q D(A_q), \text{ being minimised, or}$$

$$C(P) = \sum_q S(A_q), \text{ being maximised,}$$

[1]Actually, given the sense of step **i** of the procedure, we do not deal with the matrix of distances between individual objects, $\boldsymbol{d}$, but with the matrix of distances between clusters, $\boldsymbol{D}$, which is, at the beginning of the procedure, identical with $\boldsymbol{d}$, and then is transformed.

[2]Since we do by no means intend a survey of methods, only some selected, telling references shall be given for the methods considered. In this case—just three generic references will be mentioned: Florek, Łukaszewicz, Perkal, Steinhaus, and Zubrzycki (1956), where the origins of the so-called “single-linkage” algorithm can be found, and Lance and Williams (1966, 1967), who developed a more general theory of the agglomerative clustering procedures.

where, as can easily be seen, only the intra-cluster characteristics are accounted for. The concrete formulations of $D(.)$ and $S(.)$ in the wide variety of "k-means" algorithms, including the fuzzy-set-based ones, are such that imply the monotonicity of the optimum values of $C(P)$, $C^{opt}(P)$, in p, the globally "best" values of $C^{opt}(P)$ being inadvertently attained for $P = I$, i.e., for $p = n$.

The generic "k-means" procedure works, in general terms, in such a way that, first, p points in E_X are somehow generated (possibly randomly, or through some purposeful procedure, yielding either any points in E_X, fulfilling definite conditions, or, perhaps, a selection of the x_i's), call them x^q, $q = 1, \ldots, p$. In the next step, the objects from X are assigned to these points on the basis of minimum distance, thus forming a proper partition P, composed of p clusters. Now, for each of these p clusters we find the point, or object, which minimizes $D(A_q), q = 1, \ldots, p$. Then, we return to the assignment step. The procedure ought to terminate when two consecutive partitions coincide, although other stopping criteria are mostly applied. We shall not go into more details here, since this will be the subject of a separate section in Chap. 5.[3]

Throughout the history of cluster analysis there have been numerous ***density-based algorithms*** devised, quite differentiated as to their principles, featuring also variable popularity, due, especially, to their quite differentiated computational characteristics. All of these refer, obviously, to the idea that the clusters exist in these regions of E_X, where one can observe a high density of objects, i.e. the distances between them are relatively small and there is relatively a lot of them. Most of these methods refer to the density function, $h(x, r)$, defined as the number of objects from X, located at the distance of at most r from the point x in E_X. Two important, but also quite trivial, observations are due at this point:

(1) for $r < 1/2 \min_{i,j \in I} d(x_i, x_j)$ the density function takes just two values: 1 around the particular objects x_i, belonging to X, and 0 everywhere else, outside of the spheres around the x_i, these spheres having the diameter r;
(2) for $r \geq \max_{i,j} d(x_i, x_j)$ the density function takes the value of n, the total number of objects in X, in the region of E_X, which is equivalent to the convex hull of the set X, in other words—it is equal to n everywhere within the area spanned by the set X, and then "fades away" (its values decrease down to 0) outside of this area, that is—away from the set X.

It can already be seen that the choice of r is crucial to the formation of such a "map" of values of $h(x, r)$, which would be 'truly' telling in terms of recognition of the high density regions, corresponding to clusters, and the low density regions, corresponding to the space between clusters.

[3]Here, the seminal references are, first of all, Steinhaus (1956)—again(!), see the preceding footnote in order to appreciate the contribution of this Polish mathematician from the Lwów school of mathematics, largely founded by Stefan Banach; then there come Lloyd (1957)—soon afterwards, but similarly not 'piercing', and then Forgy (1965), Ball and Hall (1965), and MacQueen (1967). The fuzzy-set based version of the general k-means method, which became enormously popular, was formulated by Bezdek (1981).

Another approach, which refers also in some way to the local densities of objects, consists in the division of the space E_X into (usually regular and equal) cells, inside of which the objects are counted. Assume, just for the sake of illustration, that the overall division is done by dividing the continuous scales of particular variables, E_k, into equal numbers, say v, of intervals. Thus, we obtain m^v cells, and, since there are n objects, the average number of objects per cell is n/m^v. In the case of n = 10,000, m = 10, and v = 4, this average number is exactly 1. The respective techniques consist in counting the actual numbers of objects in the cells, and then concluding that the adjacent cells with the numbers of objects (much) higher than the average form clusters, while adjacent cells with (much) less objects inside them than the average—form the regions between clusters. Easier said than done, though. The algorithms, based on similar precepts, even if apparently straightforward, have not gone far in terms of application and popularity. On the other hand, they depend upon the essential choice of v, or perhaps v_k, and require quite advanced statistical testing, especially on these stages of the approach, in which it is decided whether cells, for which it can be established positively that the densities inside them are 'high', can be linked through the intermediary of other cells, for which this fact cannot be analogously established.

This stream of work encompasses quite a wide variety of approaches, starting from the early work of Raymond Tremolières (1979, 1981), who exploited in depth the properties of the density function $h(x, r)$. Both his work and the techniques of space division have not gained much success. Then, later on, a fast pragmatic technique of DBSCAN appeared (see Ester, Kriegel, Sander, & Xu, 1996, and the numerous following extensions and modifications), responding to the need for algorithms that might require just a single overview of the data set in order to form clusters and partitions, even if this means that they are less profoundly justified. This broad family of methods is still developing, with new concepts appearing, quite frequently echoing the work of Tremolières, like the currently highly popular technique, introduced by Rodriguez and Laio (2014), which has been quickly followed by numerous modifications and extensions ('fast search and find of density peaks'). Another example of developments in a similar direction are the "mountain clustering" and "subtractive clustering" algorithms, also having become quite popular (see Yager & Filev, 1994, and Chiu, 1994). Indeed, these modifications and extensions consist in a "better delineation of clusters".

From the perspective of the generic clustering problem, according to the title of this section, all of the density based algorithm emphasise the internal cohesion of clusters, while completely, or almost completely neglecting the aspect of inter-cluster dissimilarity (reduced, if at all accounted for, to the identification of "separation" of clusters).[4]

---

[4]We stop here, since his is not really a survey, but also because not so many proper clustering methods exist outside of the paradigms mentioned. Thus, for instance, the so-called ***spectral clustering*** is actually simply a dimension reduction technique, which is, in practice, coupled with the other, proper clustering methods.

The shorthand illustration, which has been provided here, referring to the most popular clustering paradigms, is, however, sufficient to note two essential ingredients of the aspect of the clustering problem that we consider here: (i) the distances and/or proximities between the objects, or, more generally, in the space E_K; (ii) a measure (criterion, objective function), allowing for the comparison of "goodness" of partitions, this measure being, in turn, based on the distances and/or proximities.

At the same time, it appears clear that the paradigms quoted here fail to reflect adequately the content of the clustering problem, and the algorithms, that they lead to, correspond to the way, in which the clustering problem is reflected.

2.3.3 *Defining a Cluster*

There exists also a "way out" of the clustering problem, consisting in proposing a constructive ("positive") *definition of a cluster*. Thus, in this perspective, a subset A of I is a cluster, if it satisfies a definite condition. We will not consider this kind of approaches, and this for two reasons:

(i) such a way of proceeding is not, at least explicitly, oriented at solving the clustering problem as is under consideration here;
(ii) in fact, the output from this kind of approach consists, as a rule, of a set of individual clusters, which:
  (ii.a) do not need to exhaust the set I, leaving some objects not clustered (usually, therefore, labelled as 'outliers'), then,
  (ii.b) they can remain in various different relations (e.g. they can overlap, or be contained one in another, etc.), and so,
  (ii.c) in many cases the (additional) problem of unique determination of the "proper" clusters arises (e.g. in the frequent cases when they are hierarchically "nested", see, e.g., Owsiński & Milczewski, 2010).[5]

A side remark is, however, perhaps due that if we knew what the proper clusters are, by the very same token we would have determined, in quite a natural manner, the "outliers", the objects or observations that "do not belong".

Actually, the same applies to the various *density-based clustering methods*, mentioned before, which are also, in fact, based on some sort of cluster definition, whether direct or implicit, referring to the locally high density of objects. This connection between the density based approaches and the ones, founded on cluster definitions, extends also to the problems encountered by both kinds of methods in determining a partition, with numerous ad hoc and poorly justified tricks used to secure obtaining of the sought solution, i.e. partition.

---

[5]The apparently highly intuitively appealing formulation: "*a cluster is a set of points x_i such that all the distances between them are smaller than between any of them and any point outside of this set*" is analysed and criticized, in particular, in Owsiński (1981, 2004a).

Now, let us also note that the formulation of the clustering problem, here considered, does not, in any manner, refer to any aspect beyond simple locations of x_i in space E_K and/or their distances and/or proximities (their concrete definitions not playing any role in the general approach, even though they may be of highest importance in the particular practical problems). Thus, for instance, we are not explicitly considering a potential probabilistic model behind the appearance of definite x_i, although, as it shall become evident, some of the purely algebraic formulae, appearing in multivariate statistics, can also be relevant here, without the probabilistic interpretation.

2.3.4 The Use of Metaheuristics

A question apart is constituted by the use of *metaheuristics* in solving of the clustering problem. With this respect let us first note that virtually all of the metaheuristics are meant to find an optimum solution for some otherwise difficult problem (e.g. multimodal objective function, highly nonlinear constraints, collinearity, etc.). This, then, presupposes the existence of the formulation of the problem in respective form, which, as said, is, in fact, missing in the standard framework of cluster analysis.

A sole exception, it seems, is constituted by the Kohonen's SOMs, the self-organising maps (see Kohonen, 2001), which, however, appear as being a close analogy of the k-means paradigm, also, therefore, in terms of the actual capacity of solving the proper clustering problem (for a discussion of this analogy, see, e.g., Bação, Lobo, & Painho, 2005, or Lindsten, Ohlsson, & Ljung, 2011).

2.3.5 The Number of Clusters

The failure of the most popular—if not simply all—approaches, used for clustering purposes to address the entire formulation of the generic problem of clustering entails also one of the fundamental issues of cluster analysis: *how many clusters are there*? The progressive merger (hierarchical clustering) schemes leave the question aside at all, while the k-means techniques imply the "global optimum" in the form of all objects being separate clusters each. The density-based approaches determine some number of clusters, potentially forming a partition, but the implicit cluster definitions, used by them, are not only subject to the choice of respective parameters ("neighbourhood distance", "neighbour number", "density threshold", etc.), but also rely on only one side of the clustering problem, i.e. the inner cluster coherence.

As said, this calls for the application of external, usually derived from statistics, criteria that would indicate the quality of different partitions, including also the potential indication of the number of clusters involved. The popular statistical packages offer "validation" of partitions with these criteria, but their number, which is still increasing, has gone by now well beyond 40 (consult, for instance, among many papers on this subject, Vendramin, Campello, & Hruschka, 2010, or the materials, concerning the R package).[6]

2.3.6 *The Shapes of Clusters*

Another key problem of cluster analysis is that of *cluster shapes* (are we capable of identifying clusters of various shapes?). This problem concerns, of course, primarily the shapes significantly differing from spherical ones, complicated, with non-trivial interrelations between clusters, as this is illustrated by the instance of Fig. 2.3.

In this example it is easy to see, as well, that the issue of shape may be closely associated with that of the number of clusters. In fact, for complex cluster shapes the question "is this a [single] cluster?" arises in quite a natural manner (are there really four clusters in Fig. 2.3? or perhaps just one?).

The issue of cluster shape shall not be tackled in this volume, and that for two reasons:

(1) the approach that we propose here can be applied to a variety of methodological precepts that can be used to identify partitions composed of clusters having various types of shapes;
(2) the issue of shape is less general than that of the number of clusters, and in many instances the methods, which are aimed at specific cluster shapes, may therefore be quite specialised (e.g. related to definite cluster interpretation), with less attention being paid to other properties of the methods, like numerical efficiency, validity of results, or recovering of the cluster number.

2.3.7 *Why not the Exhaustive Search?*

To close these extensive introductory considerations let us emphasise that, indeed, notwithstanding the issue of the "appropriate" formulation of the generic clustering problem, definite algorithmic inventions are necessary in solving the clustering problems, even of quite moderate dimensions (say—thousands of objects), because complete enumeration is, definitely out of question.

---

[6]Note that in this context we refer only to those of the partitioning or clustering criteria alluded to that are called "internal" (see, e.g., Rendón, Abundez, Arizmendi, & Quiroz, 2011), since the ones called "external" actually verify the classification capabilities of the respective methods, and do not address the clustering performance as such.

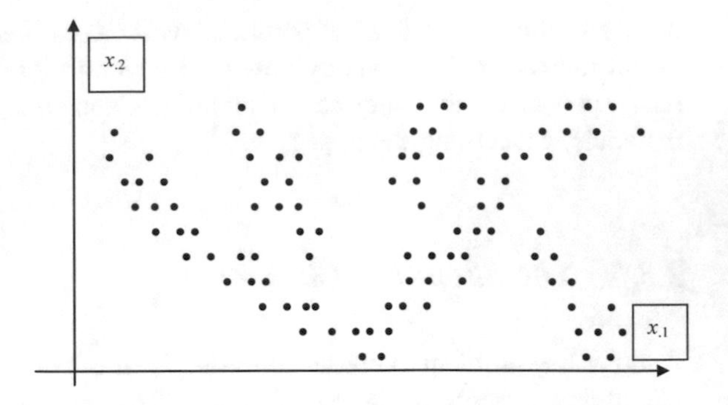

Fig. 2.3 An example of a set of objects featuring complex clusters shapes

Namely, the number of partitions of the set *I*, composed of *n* elements, just as we understand partitions here, is the Bell number B_n (see, for instance, Rota, 1964). Bell numbers form a recursive sequence, with $B_0 = B_1 = 1$, $B_2 = 2$, $B_3 = 5$, $B_4 = 15$, etc., the recursion formula being

$$B_{n+1} = \sum_{\nu=0}^{n} \binom{n}{\nu} B_\nu.$$

This formula yields truly formidable numbers of partitions. And so, while $B_5 = 52$, and $B_6 = 203$, if we go up to just $n = 15$, we get $B_{15} = 1{,}382{,}958{,}545$, and for $n = 19$ we get $B_{19} = 5{,}832{,}742{,}205{,}057$, these being at most equivalent to toy examples of data sets. To illustrate more strongly the scale of this issue, let us say that the biggest prime Bell number, that was known as of the year 2006, was B_{2841}, approximately equal to $9.30740105 \times 10^{6538}$, and this for $n = 2841$, that is—still quite a modest magnitude of the data set, while we have to deal in practice with many thousands or even millions of objects.

Thus, even if we can eliminate relatively simply quite a proportion of the partitions from our considerations, on the basis of some preliminary assessments (although this is, in general, not certain that we shall be able to do so), the very dimensions of the problem call definitely for algorithmic inventions.

2.3.8 *On Clustering Validity Indices*

We have already several times mentioned the existence of a truly high number of partitioning or clustering indices or criteria, which are quite often used. The need to use them results from the fact that, in general, the existing clustering methods fail to adequately represent the entire generic formulation of the clustering problem (the two sides of it) and therefore give rise to the necessity of evaluation of the results provided by them. Actually, the examples of references already recalled, like Rendón et al. (2011), Vendramin et al. (2010), or the materials, concerning the R package, and many others, demonstrate that—in turn—the very number of the indices or criteria results from the lack of clear "prescription" as to the quality of the partition obtained (once we do not know how to represent adequately the clustering problem, we do not also know how to measure the quality of the clusterings). The respective references are, namely, most often the comparative studies, involving the most popular indices, such as Calinski-Harabasz, Davies-Bouldin, Dunn, variations of the silhouette index, Akaike's information criterion, Bayesian information criterion—both of the latter in appropriate adopted forms, and so on.

Yet, the comparison can hardly be carried out against a well-defined "objective" gold standard, since there is no such thing in clustering analysis. Hence, the comparisons stop, as a rule, at the point of indicating the similarities and differences among the indices versus a number of well known, benchmark data, for which it is often supposed that we know the "true" partition (while it is by no means obvious that such a partition, even if very sound and justified, has anything to do with the working of the clustering procedures).

Actually, the very names of these criteria functions or statistical yardsticks tend often to be misleading. Thus, for instance, the name like "clustering validity index", obviously suggests that we know sufficiently well the statistical properties that a clustering ought to satisfy.[7] These names are, again, a separate subject of discussion (e.g. why not "clustering quality criterion"?).

Finally, however, let us admit that many of the indices used in fact do embed the representation of the two sides of the generic clustering problem, even if they are used to evaluate the (results of the) methods that fail to do so. That is why at the end of Chap. 5, in Sect. 5.4, we provide a limited comparison of the precepts of the bi-partial approach with some of the indices of clustering quality.

References

Bação, F., Lobo, V., & Painho, M. (2005). Self-organizing maps as substitutes for k-means clustering. In V. S. Sunderam et al (Eds.), *ICCS 2005*, (LNCS 3516, pp. 476–483).

Ball, G., & Hall D. (1965). ISODATA, a novel method of data analysis and pattern classification. Technical report NTIS AD 699616. Stanford Research Institute, Stanford, CA.

Bezdek, J. C. (1981). *Pattern recognition with fuzzy objective function algorithms*. New York: Plenum Press.

Chiu, S. L. (1994). Fuzzy model identification based on cluster estimation. *Journal of Intelligent & fuzzy systems, 2*, 267–278.

Ester, M., Kriegel, H.-P., Sander, J., & Xu, X.-W. (1996). A density-based algorithm for discovering clusters in large spatial databases with noise. In E. Simondis, J. Han, U. M. Fayyad. (Eds.), *Proceeding of the Second International Conference on Knowledge Discovery and Data Mining (KDD-96)* (pp 226–231). AAAI Press.

Florek, K., Łukaszewicz, J., Perkal, J., Steinhaus, H., & Zubrzycki, S. (1956). Taksonomia Wrocławska (The Wrocław Taxonomy; in Polish). *Przegląd Antropologiczny*, 17.

[7]This supposition is, of course, true, when we deal with a definite, very narrow class of data sets, e.g. we can assume all clusters correspond to some Gaussian distribution functions.

Forgy, E. W. (1965). Cluster analysis of multivariate data: efficiency versus interpretability of classifications. Biometric Society Meeting, Riverside, California, 1965. Abstract in *Biometrics* (1965) 21, 768.

Kohonen, T. (2001). *Self-organizing maps*. Berlin-Heidelberg: Springer.

Lance, G. N., & Williams, W. T. (1966). A generalized sorting strategy for computer classifications. *Nature, 212,* 218.

Lance, G. N., & Williams, W. T. (1967). A general theory of classification sorting strategies. 1. *Hierarchical Systems. The Computer Journal, 9,* 373–380.

Lindsten, F., Ohlsson, H., & Ljung, L. (2011). Just relax and come clustering! A convexification of k-means clustering. Technical Report, Automatic Control, Linköping University, LiTH-ISY-R-2992.

Lloyd, S. P. (1957). Least squares quantization in PCM. Bell Telephone Labs Memorandum, Murray Hill, NJ; reprinted in *IEEE Transactions Information Theory*, IT-28 (1982), *2*, 129–137.

MacQueen, J. (1967) Some methods for classification and analysis of multivariate observations. In L. M. LeCam, J. Neyman, (Eds.), *Proceedings* 5th *Berkeley Symposium on Mathematical Statistics and Probability* 1965/66 (vol. **I**, pp. 281–297). University of California Press, Berkeley.

Owsiński, J. W. (1981). Intuition versus formalization: local and global criteria of grouping. *Control and Cybernetics, 10*(1–2), 73–88.

Owsiński, J.W. (2004a). Group opinion structure: The ideal structures, their relevance and effective use. In D. Baier & K.-D. Wernecke, (Eds.), *Innovations in Classification, Data Science, and Information Systems. Proceeding 27th Annual GfKl Conference, University of Cottbus, March 12-14, 2003* (pp. 471–481), Springer, Heidelberg-Berlin.

Owsiński, J. W., & Milczewski, M. (2010). Rekursja w problemie regionalizacji (Recursion in the regionalisation problem; in Polish). In J. W. Owsiński, (Ed.) *Analiza systemów przestrzennych. Wybrane zagadnienia. Badania Systemowe* (vol. 6, pp. 47–587). Instytut Badań Systemowych PAN, Warszawa.

Rendón, E., Abundez, I., Arizmendi, A., & Quiroz, E. M. (2011). Internal versus external cluster validation indexes. *International Journal of Computers and Communications, 5*(1), 27–34.

Rodriguez, A., & Laio, A. (2014). Clustering by fast search and find of density peaks. *Science, 322,* 1492.

Rota, G.-C. (1964). The number of partitions of a set. *The American Mathematical Monthly, 71*(5), 498–504.

Steinhaus, H. (1956). Sur la division des corps matériels en parties. *Bulletin de l'Academie Polonaise des Sciences, IV (C1.III)*, 801–804.

Tremolières, R. (1979). The percolation method for an efficient grouping of data. *Pattern Recognition*, 11.

Tremolières, R. (1981). *Introduction aux fonctions de densité d`inertie* (p. 234). IAE: Université Aix-Marseille, WP.

Vendramin, L., Campello, R. J. G. B., & Hruschka, E. R. (2010). Relative clustering validity criteria: A comparative overview. *Wiley InterScience*. https://doi.org/10.1002/sam.10080.

Yager, R. R., & Filev, D. P. (1994). Approximate clustering via the mountain method. *IEEE Transactions on Systems, Man, and Cybernetics, 24,* 1279–1284.

Chapter 3
The General Formulation of the Objective Function

3.1 The Formulation

The objective function, which is the main subject of this volume, has the following general form:

$$Q_S^D(P) = C_S(P) + C^D(P) \tag{3.1}$$

where $C_S(P)$ is the measure of quality ("goodness") of partition P with respect to proximity of objects contained inside particular clusters, forming P (that is: with respect to internal proximity of clusters), while $C^D(P)$ is the measure of quality of partition P with respect to distances between objects in different clusters (distances between clusters). $Q_S^D(P)$ is, of course, maximised over the space of partitions E_P.

As a kind of a "dual", we can equally well define

$$Q_D^S(P) = C^S(P) + C_D(P) \tag{3.2}$$

in which $C^S(P)$ is the measure of quality of partition P with respect to proximity between the objects located in different clusters (between different clusters) and $C_D(P)$ is the measure of quality of partition P with respect to distances between objects located in the same clusters (distances characterising individual clusters). The function $Q_D^S(P)$ is, naturally, minimised over E_P.

For the sake of a trivial illustration, let us remind of the k-means-type formulation of the objective function, mentioned already, equivalent, actually, to formulation of some $C_S(P)$ or $C_D(P)$, namely:

$$C_D(P) = \sum_q D(A_q), \text{ minimised, or}$$

$$C_S(P) = \sum_q S(A_q), \text{ maximised.}$$

J. W. Owsiński, *Data Analysis in Bi-partial Perspective: Clustering and Beyond*, Studies in Computational Intelligence 818,
https://doi.org/10.1007/978-3-030-13389-4_3

Indeed, these formulations indicate clearly that there is a basic difference between the formulations (3.1) and (3.2) on the one hand, and those proper for the k-means-type algorithms on the other. The latter obviously account for just one "side" of the clustering problem, while neglecting the other.[1]

3.2 Some Explanations and the Leading Example

Both of the "dual" forms of the general objective function are meant to reflect the essence of the clustering problem, i.e. "similarity inside clusters and dissimilarity among clusters". That is why both are composed of two parts, one responsible for the intra-cluster proximity (similarity), and the other one—for inter-cluster distances (dissimilarities). The forms (3.1) and (3.2) explain, therefore, the name appearing in the title of the present volume: the "bi-partial objective function".

Although for the adequate representation of the clustering problem the two components could be put together ("…and…") in some other way than in (3.1) and (3.2) (like, e.g., through the product of these two parts), we shall stick to the simple sum, for several reasons, which will get clearer further on. For a while let us only emphasise that this is perhaps the simplest manner, in which they can be put together in order to adequately reflect the essence of the clustering problem.

It is assumed that the general forms here presented are turned into concrete implementations, so that their values can be practically calculated for the given data sets, i.e. objects X_I and for their partitions P.

The four functions C, appearing in formulae (3.1) and (3.2), which are meant to reflect the one-sided quality of respective partitions, are, naturally, based upon the x_i, d_{ij}, s_{ij}, $D(A_q,A_{q'})$, $D(A_q)$, $S(A_q,A_{q'})$, $S(A_q)$, with the selection of these entities and the forms of respective functions depending upon the concrete implementation, which, in turn, supposedly depends upon the nature of the problem at hand. The existing methods offer already quite a choice of these formulations, including C's. We shall provide further on a number of examples of implementations of Q, based on various concepts as to what a good partition may be.

---

[1]At this point a remark is due on the sometimes voiced argument that by minimising the function like $C_D(P)$ for k-means-type algorithms, one maximises, at the same time, the respective inter-cluster distance measure, and so, by this means, the "bi-partiality" is secured. Such an argument is, of course, void, since the "other side" has to be maximised simultaneously, and not "by implication", for otherwise, as already noted, we end up with the trivial solution for P, namely $p = n$.

For the sake of a more detailed illustration and analysis let us introduce here the concrete form of the general objective function that will be most frequently referred to further on,[2] possibly the simplest and the most intuitive one, namely

$$\begin{aligned} Q_S^D(P) &= C_S(P) + C^D(P) \\ &= \sum_q \sum_{i<j\in Aq} s_{ij} + \sum_q \sum_{q' > q} \sum_{i\in Aq} \sum_{j\in Aq'} d_{ij}. \end{aligned} \quad (3.3)$$

Assuming that d_{ij} are defined for a given problem in some "natural" manner, we can define s_{ij} as, e.g.,

$$s(d) = d^{\max} - d, \quad (3.4)$$

where $d^{\max} = \max_{i,j} d_{ij}$ is the diameter of the set of objects, X, or, equivalently, after normalisation, i.e., when $d_{ij} \in [0,1]\ \forall\ i,j\in I$,

$$s(d) = 1 - d. \quad (3.5)$$

In formulation of the objective function (3.3) it is easily seen that if the partition P^* is composed of just one all-embracing cluster, $p = 1$, $A_1 = I$, then

$$Q_S^D(P^*) = C_S(P^*) = \sum_{i<j\in I} s_{ij} = S(I), \quad (3.6)$$

which, for the definition of $s(.)$ as in (3.4), turns into

$$Q_S^D(P^*) = S(I) = d^{\max} n(n-1)/2 - D(I).$$

On the other hand, if the partition P^{**} is composed of all the objects being separate clusters, $p = n$, $A_q = q$, then

$$Q_S^D(P^{**}) = C^D(P^{**}) = \sum_{i<j\in I'} d_{ij} = D(I). \quad (3.7)$$

Comparison of (3.6) and (3.7) may give rise to application of yet another definition of $s(.)$, so as to secure that $Q_S^D(P^*) = Q_S^D(P^{**})$, the equality of the objective function values at the two extremes, regarding the partitions. The function $s(.)$, which satisfies this condition, and at the same time the reasonable condition of $s^{\min} = d^{\min}$ (generally stronger than the necessary $s^{\min} \geq 0$), has the form

[2]This particular ("leading") concrete formulation is considered at greater length, due to its intuitive appeal, significance regarding the possibility of constructing the sub-optimising algorithm and some essential "historical" references, later on in this volume, especially in Sect. 5.1, in Chap. 6 and at the beginning of Chap. 7.

$$s(d) = d^{average}\frac{d^{\max} - d^{\min}}{d^{\max} - d^{average}} - \frac{d^{average} - d^{\min}}{d^{\max} - d^{average}}d. \quad (3.8)$$

On the other hand, one could also reason for the use of the function *s*(.) that preserves the range of values of *d*, i.e. $s^{\max} - s^{\min} = d^{\max} - d^{\min}$, requiring, again, additionally, $s^{\min} = d^{\min}$, which is equivalent to a simple definition

$$s(d) = d^{\min} + d^{\max} - d. \quad (3.9)$$

There may also exist other kinds of arguments for choosing both the definition of *d*(.,.) and the transformation *s*(*d*) (or vice versa), associated with the given kind of problem or nature of the data set—e.g. the particular distribution of the values of distances, calling for an appropriate transformation *s*(*d*) in terms of properties of the resulting distribution of proximities.

With these exemplary definitions we are now able of providing a simple numerical illustration, showing some of the essential properties of the bi-partial objective function. The respective two-dimensional "data set" is provided in Fig. 3.1. For this set of points ("objects") we show the sequence of values of the objective function (3.3) (with (3.8)) corresponding to the following sequence of nine nested partitions, with the objects (points) being generally numbered from down left to right:

(1) {1}{2}{3}{4}{5}{6}{7}{8}{9}
(2) {1, 2} {3}{4}{5}{6}{7}{8}{9}
(3) {1, 2} {3}{4, 5} {6}{7}{8}{9}
(4) {1, 2} {3}{4, 5} {6}{7}{8, 9}
(5) {1, 2, 3} {4, 5} {6}{7}{8, 9}
(6) {1, 2, 3} {4, 5, 6} {7}{8, 9}
(7) {1, 2, 3} {4, 5, 6} {7, 8, 9}
(8) {1, 2, 3, 4, 5, 6} {7, 8, 9}
(9) {1, 2, 3, 4, 5, 6, 7, 8, 9}.

Actually, such a sequence, consisting of (nested) partitions, formed by consecutively aggregating clusters, could be considered typical for (in fact: could have resulted from) the one of the classical hierarchical (agglomerative) algorithms of cluster analysis.

The shape of the function (3.3) for this sequence and for the definition of the transformation *s*(*d*) according to (3.8) is shown in Fig. 3.2. Since the function (3.3) is being maximised, the function obviously indicates—for this sequence—partition no. 8, i.e. composed of just two clusters, as the best one.

If this result is not fully in line with the intuition of the Reader, then the case of Figs. 3.3 and 3.4 might perhaps be.

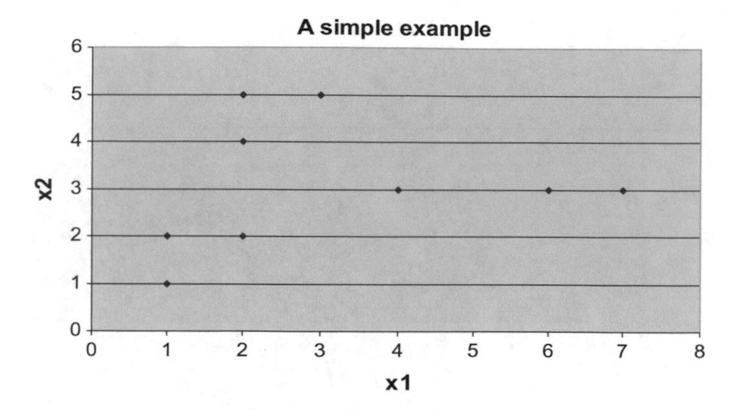

Fig. 3.1 A simple example of a data set

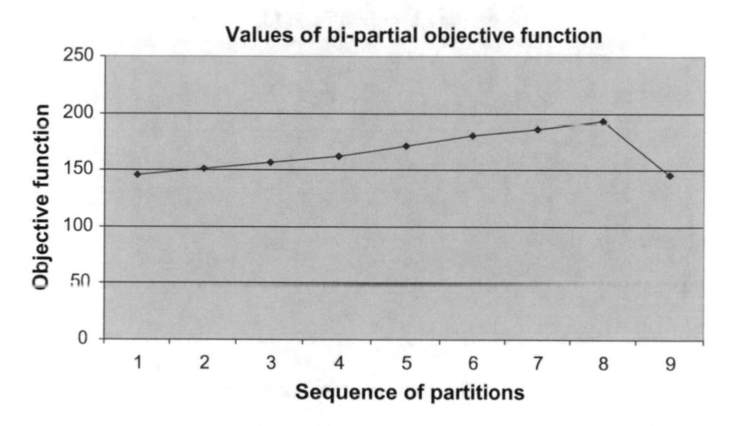

Fig. 3.2 Values of the bi-partial objective function for the data of Fig. 3.1 and the sequence of partitions given in the text

Figure 3.4 shows that for this case with the same sequence of partitions, but a slightly different data set—partition no. 7, composed of three clusters, comes out as the best one.

For a Reader, insisting, upon the (quite understandable) intuitive precepts, that object (point) no. 7, located apparently between the three bigger groups of objects, ought to form a separate cluster, we show the example of Figs. 3.5 and 3.6, where the data points have again been slightly modified, with preservation of the general character, but, even more importantly, the definition of $d \rightarrow s$ has been changed to $s(d) = 7 - 0.6d$, while, in this case, the characteristic values of distances are: $d^{\text{min}} = 1$, $d^{\text{max}} = 11$, and $d^{\text{average}} = 6.16(6)$. Now, an attentive Reader might rightly suspect that this amounts to manipulating the perception of the scale, regarding our academic data set.

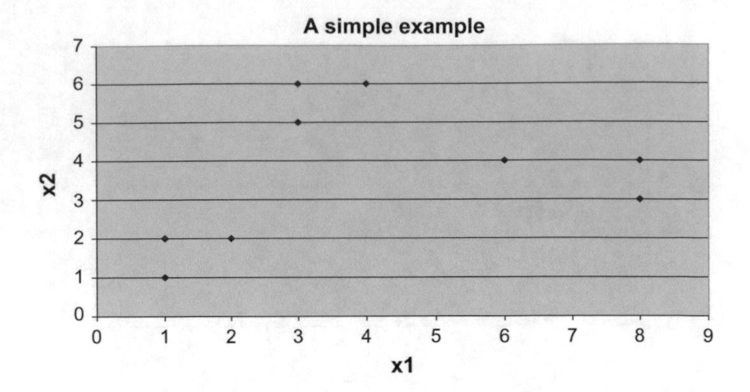

Fig. 3.3 Another simple example of a data set, similar to that of Fig. 3.1

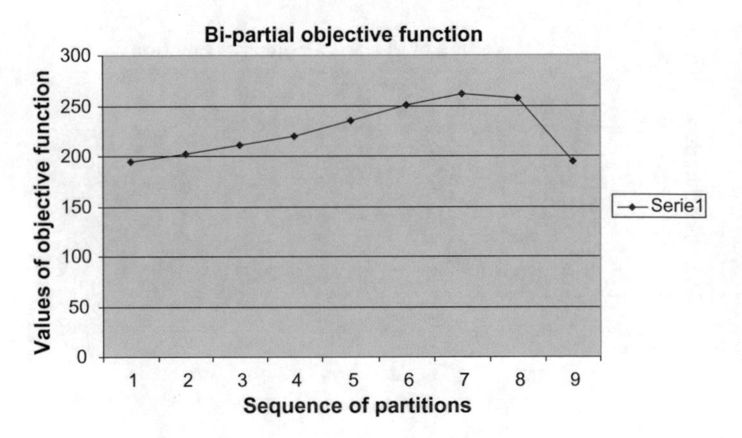

Fig. 3.4 Values of the bi-partial objective function for the data of Fig. 3.3 and the sequence of partitions given in the text

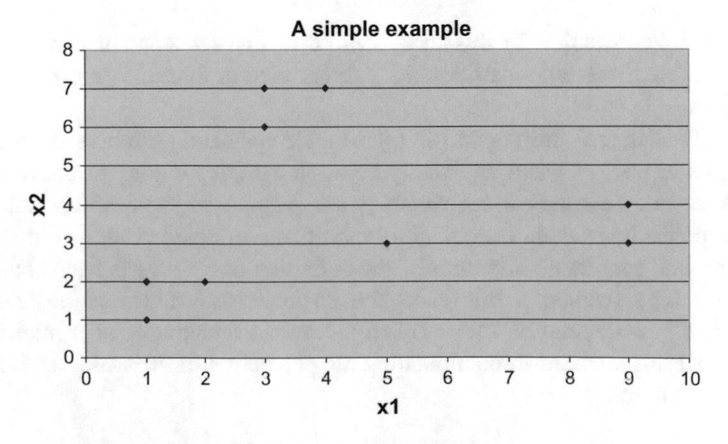

Fig. 3.5 Yet another—similar—simple example of a data set

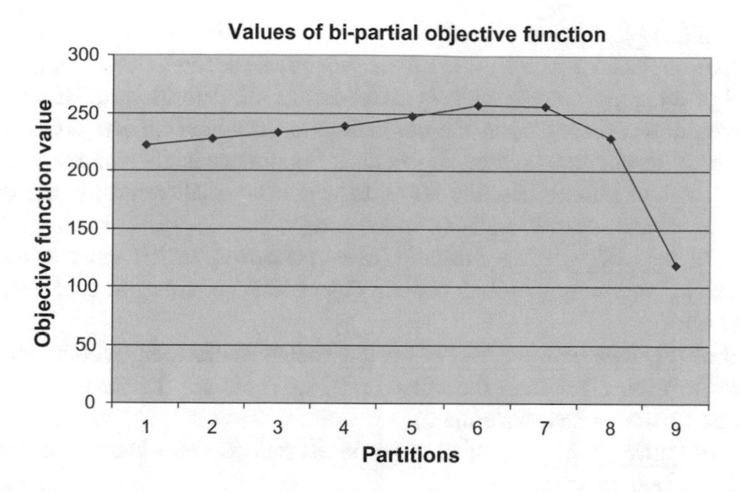

Fig. 3.6 The values of the bi-partial objective function for the data of Fig. 3.5 and specifications as in the respective text

As can be seen from Fig. 3.6, now the maximum is attained for partition no. 7, as "wished", but, given the specifications provided above, it took some "forcing" to get this solution, which, however, some may find intuitively plausible, or even "correct".

3.3 A Broader View: The Levels of Perception and the Issue of Scale

We shall not, at this point yet, introduce any assumptions concerning the properties of C's, since they can be so very diversified that postulating or requiring such properties would always be a limitation on the potential choice of formulations. Some relatively mild conditions will be introduced later on, while here it suffices to state that the C's should "correctly" reflect the quality of partitions in the sense of either inter- or intra-cluster perspective.

Still, in connection with the above and the illustrative example, in which we explicitly referred to "intuitions" and to "scale", let us note that all the formulations that can be incorporated in the general framework here considered, are based on some sort of "perception", having a clear intuitive and interpretable appeal, leading to concrete, but highly diversified, mathematical expressions. The perceptions, which can stem either from the domain-specific knowledge, or from convictions, related to a broader overview of the data analysis area, including formal precepts, refer to several distinct "levels of perception", i.e.:

– level of a variable (definition of a variable, its scale, expression of values; space E_k),

- level of an object (what is an object? how do we distinguish objects? how do we compare objects? space E_K); are there "identical objects" ("repetitions")?
- level of relations between objects (here, first of all: definitions of distances and/or proximities, based upon the conviction as to which objects are closer and which are farther away); here, of foremost importance is the relation of scales of distances and proximities, the issue, to which we shall return; is the distance from an object to itself really (always) zero?
- level of sets of objects (clusters) (here, primarily: which clusters are more "compact", meaning that they contain objects that are closer to each other, than other ones),
- level of relations between sets of objects (between clusters) (which clusters are closer to each other than the other ones? i.e. how are distances and/or proximities between clusters defined);
- level of entire partitions (which of partitions reflects better the mutual positions of the objects in space?).

At each of these levels of perception definite intuitive concepts exist, justifying the use of concrete mathematical conditions and formulations. It is from these various concepts, referring to various levels of perceptions, that different methods and algorithms, including those of cluster analysis, arise.

It is, of course, not necessary to build the functions C on the basis of the entire hierarchy of these levels of perception, at least explicitly. In most cases some of these levels are in a way omitted, like when $C_D(P) = \max_q D(A_q)$, with, in turn, $D(A_q) = \max\{d_{ij}|i,j \in A_q\}$, so that $C_D(P)$ is represented by just one distance value.

There is, however, an important aspect in the existence and role of the particular levels of perception. Namely, the convictions as to the "correct" or "adequate" representations at different levels need not be consistent, or a special care must be taken in order to keep them consistent. In order to illustrate this aspect let us simply note that even in the case, considered above, of an easy formulation of $C_D(P)$ it was necessary to define and use notions from a number of distinct levels of perception.

Against this background let us forward some observations, concerning the perception levels, which were made explicit in the formulae (3.3–3.8), for instance. Thus, first, it appears reasonable to have the potential contribution from both parts of $Q^D_S(P)$ or $Q^S_D(P)$ balanced in some way. This may mean, for example, that the number of elements in these two parts is kept roughly the same or that the sum of the numbers of elements is constant (like in (3.3), where it is always equal to $n(n - 1)/2$). Another, and simpler, way of imposing this requirement might be to have $S(I) = D(I)$.

A different observation with regard to the basic formulation of the bi-partial objective function, and its implementations, like that of (3.3), refers to the question:

why should we insist on the formulation of this objective function in terms of distances and proximities at the same time, especially as we explicitly define the transformations $d \leftrightarrow s$*?*

Is the formulation in terms of, say, d alone, not only simpler and more elegant (not involving an additional aspect that can be suspected to introduce additional subjectivity or arbitraryness), but also easier to treat?

For purposes of illustration, let us reformulate (3.3) by employing (3.4) so as to get rid of proximities s:

$$\begin{aligned} Q_S^D(P) &= C_S(P) + C^D(P) \\ &= \sum_q \sum_{i<j \in Aq} \left(d^{\max} - d_{ij}\right) + \sum_q \sum_{q' > q} \sum_{i \in Aq} \sum_{j \in Aq'} d_{ij} \\ &= \sum_q \sum_{q' > q} \sum_{i \in Aq} \sum_{j \in Aq'} d_{ij} - \sum_q \sum_{i<j \in Aq} d_{ij} \\ &+ \sum_q 1/2 \mathrm{card} A_q \left(\mathrm{card} A_q - 1\right). \end{aligned} \tag{3.10}$$

Apparently, by this transformation we have not gained much—indeed nothing—especially in terms of the potential algorithmic facility, and we definitely lost the intuitive appeal of the original formulation (now, in addition to distances, we deal also with a function of cardinalities of clusters!).

Actually, the form of the objective function we propose is based on both the intuitive appeal of the basic formulation and (of) the associated properties, which shall be discussed further on.

There is, though, one more important aspect, associated with the explicit use of both distances and proximities, having simultaneously intuitive and pragmatic significance, namely that of *scale*, that we have already alluded to.

The illustration to the point is provided in Fig. 3.7. Thus, consider the case of Fig. 3.7a, and then compare it with the one of Fig. 3.7b. It should be emphasised that these two figures present the set of objects on a plane that can be described by the same values of coordinates and the same values of pairwise distances.

What—quite intuitively—differentiates these two cases, is exactly the *perception of scale*, suggested by the frames, supposedly representing the feasible ranges of values of respective variables. For a person, who performs the analysis, this perception, provided by the known ranges of variables values, ought to come as quite natural.

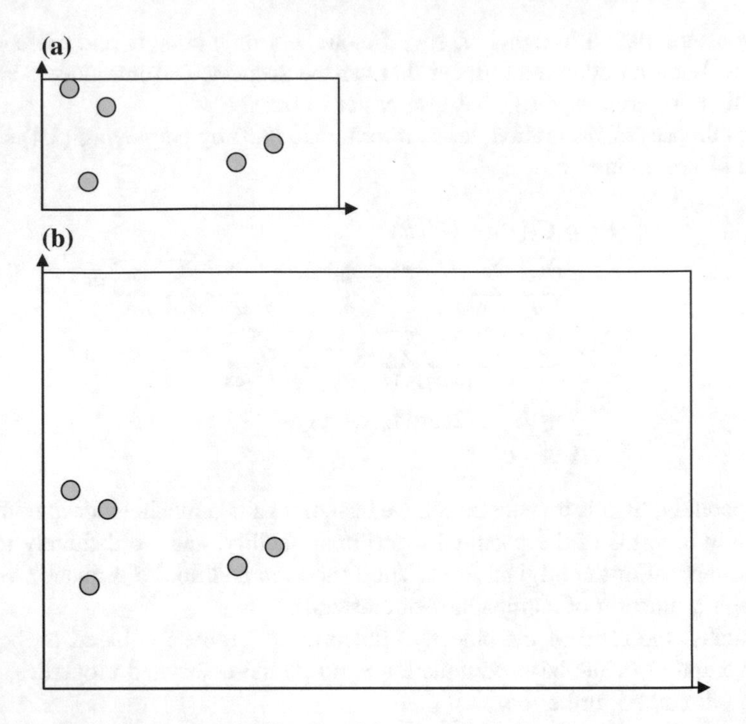

Fig. 3.7 An illustration for the significance of the perception of scale

The pertinent question that we wish to ask here is: are we dealing with two (three?) clusters or just one cluster? The answer depends exactly upon the assumed scale, shown in Fig. 3.7a, b in terms of surrounding frames. This scale can be naturally incorporated into the formulation of the clustering problem according to the bi-partial objective function through the use of appropriate explicit transformation $s \leftrightarrow d$. Such an operation would be quite artificial for the formulation based uniquely on distances d. Thus, we obtain the instrument for modelling of the aspect of scale, an inherent aspect of the clustering problem in general.[3]

[3]Conform to the proposition that the entire problem of clustering is formulated according to the levels of perception, also the aspect of scale appears at such various levels. Here it was illustrated for the level of objects, but it might be also reasonable to consider it for the level of entire clusters, distances and proximities d and s being appropriately replaced by the functions D and S.

Chapter 4
Formulations and Rationales for Other Problems in Data Analysis

We shall now comment upon some other problems, considered in the broadly conceived domain of data analysis, as seen in the perspective of the bi-partial objective function. We shall show the applicability of the concept of bi-partial objective function to these (and, indeed, yet other) problems with, whenever appropriate, illustrations for the potential form of the respective objective function.

4.1 Categorisation

The task of categorisation, often treated as an element of the initial processing of data for other, more in-depth problems, consists in splitting of the space of values of a variable k, i.e. the set E_k, into subsets, denoted $C_{kc}, c = 1, \ldots, c^{\max}(k)$, these subsets corresponding to "categories", referred to through indices c, and used in further procedure instead of values $x_{ik} \in C_{kc}$. It is standard to assume $C_{kc} \cap C_{kc'} = \emptyset$ for all $c \neq c'$, and also, quite naturally, $\cup_c C_{kc} = E_k$. The purpose, as said, is to have the "categories" C_{kc}, and hence the indices c, to replace the original values x_{ik} in such a way as to possibly well represent the distribution of values of x_{ik} within the space E_k.

This description of the task of categorisation implies that, under certain conditions (e.g. how are the particular values x_{ik} characterised, i.e. what is the "basis for categorisation"), it could be seen as a uni-dimensional clustering, with categories corresponding to clusters, and all the consequences of this implication, pertinent to clustering, holding. Even though the requirement of "representation" seems to make this task a bit more specific than the general problem of clustering, unless the "representation" is well defined (see the other problems here considered!), we might assume that clustering is a proper model for categorisation.

J. W. Owsiński, *Data Analysis in Bi-partial Perspective: Clustering and Beyond*, Studies in Computational Intelligence 818,
https://doi.org/10.1007/978-3-030-13389-4_4

There exist several results, dealing directly with the problem of categorisation (see, for instance, Gan, Ma, & Wu, 2007, pp. 30ff.), as well as algorithms, but we shall not discuss them here.

To start with consideration of this problem, let us indicate that in the analysis of categorisation it is standard to treat the values $x_{ik} \in E_k$ as ordered. For univariate analysis oriented at categorisation this is, of course, fully justified. A very simple instance of a set of ordered values along one variable is shown in Fig. 4.1.

In the illustration of Fig. 4.1, the space of variable values appears as denoted by E^*_k, to indicate that it is transformed by ordering of values of x_{ik}. A simplistic question to ask for the case thus illustrated, given the gap between the values ordered as 7th and 8th, would be: should there be two or more categories? or as many as there are different variable values (the latter choice meaning that we actually give up categorisation altogether)?

All this implies that we are here in exactly the same position as in the general clustering problem, including the issue of the number of clusters (here: categories). Indeed, we can define d_{ij} simply as $d_{ij} = \left|x_{ik} - x_{jk}\right|$ and construct functions $D(.,.)$, $S(.)$, and then $C^D(P)$ and $C_S(P)$, as well as $Q_S^D(P)$, on this basis. In this way we would get the categories that would be internally possibly similar ("compact"), while being possibly strongly differentiated among themselves. If this is accepted as corresponding to appropriate "representation", then the task of categorisation could be solved with the use of the bi-partial objective function.

Given that we deal with (upward) ordered values of x_{ik}, the subsets ("clusters") C_{kc} will be defined as subsequences of values x_{ik} through the respective "bordering" values in the following manner:

$$C_{kc} = \left[x_{i1(c)k}, x_{i2(c)k}\right],$$

with, of course,

$$x_{i1(c)k} \leq x_{i2(c)k}, \quad \text{and} \quad x_{i2(c)k} \leq x_{i1(c+1)k}.$$

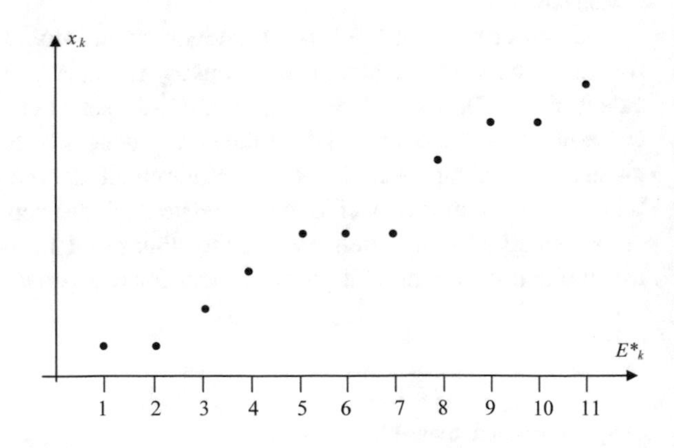

Fig. 4.1 An illustration of data for the task of categorisation

The (ordered) set (i.e. a subsequence) of indices i of objects (measurements) forming the category C_{kc} shall be denoted $I(kc)$.

The bi-partial function might, then, for this problem, take the following form, which is maximised over the partitions into C_{kc}:

$$Q_S^D(P) = C^D(P) + C_S(P) = \sum_c d\big(x_{i2(c)k}, x_{i1(c+1)k}\big) + \sum_c \sum_{i \in I(kc)} s\big(d_{i,i+1}\big) \quad (4.1)$$

with $s(.)$ being an appropriately, for this case, defined proximity function, e.g.

$$s(d) = \max_i d_{i,i+1} - d = d_i^{\max} - d. \quad (4.2)$$

It can be seen that also here we aim to maximise this function without any assumptions on the number of categories (clusters), $c^{\max}(k)$, which should result from maximisation of $Q_S^D(P)$.

Similarly as in the case of objective function (3.3) we might propose to cast (4.1) into the formulation involving only distances d. We then get

$$\begin{aligned} Q_S^D(P) &= C^D(P) + C_S(P) \\ &= \sum_c d\big(x_{i2(c)k}, x_{i1(c+1)k}\big) + \sum_c \sum_{i \in I(kc)} \big(d_i^{\max} - d_{i,i+1}\big) \\ &= \sum_c d\big(x_{i2(c)k}, x_{i1(c+1)k}\big) - \sum_c \sum_{i \in I(kc)} d_{i,i+1} + n d_i^{\max}, \end{aligned} \quad (4.3)$$

which, again, as in the case of clustering, represents no practical advantage over the formulation (4.1).

Now, like in the case of the function (3.3), we shall illustrate the sense of the objective function (4.1), along with (4.2), for the extremely simple case of Fig. 4.1, in which we treat the values 1 through 11, shown on the horizontal axis (indices of objects), as genuine values of the variable considered, in order not to mislead the intuition of the Reader. We show below the values of (4.1) for a selected sequence of partitions, similar to that for the example from the preceding section.

Partitions (categorisations)	*Objective function values*
$\{1\}\{2\}\{3\}\{4\}\{5\}\{6\}\{7\}\{8\}\{9\}\{10\}\{11\}$	7
$\{1,2\}\ \{3\}\{4\}\ \{5,6,7\}\ \{8\}\ \{9,10\}\{11\}$	15
$\{1,2,3\}\ \{4,5,6,7\}\ \{8,9,10,11\}$	15
$\{1,2,3,4,5,6,7\}\ \{8,9,10,11\}$	15
$\{1,2,3,4,5,6,7,8,9,10,11\}$	13

This result, given that only few integer values of the variable are involved and a number of pairs of objects are equidistant, is intuitively "correct". Indeed, it is hard to say which one, of the three "intermediate" partitions, is the most appropriate one for this case.

There may, however, in general, exist some special (but also quite natural) requirements, set on the categorisation task, like the stipulation of equal cardinality of C_{kc} for all c (i.e. equal width of intervals in continuous case), or the maximum allowed number of categories, that may make (more) difficult the reference to clustering in general, and to the bi-partial objective function in particular. In fact, the following section is devoted to this kind of problem.

4.2 Optimum Histogram

This problem, very similar to the preceding one, is: for a univariate set of data, keeping previous notation, i.e. $x_{ik} \in E_k$, we wish to determine, like before, the subsets of E_k, denoted C_{kc}, so that the histogram, based on $\{C_{ck}\}$, i.e. the set of pairs $\{n_{ck}, C_{ck}\}, c = 1, \ldots, c^{\max}(k)$, where n_{ck} is the number of objects x_i such that $x_{ik} \in C_{ck}$, represents possibly faithfully the actual distribution of values of this variable. In doing this, we assume that $c^{\max}(k) \ll n$. The purpose is to provide a shorthand, synthesising image of the actual distribution of values.

For a simple illustration of this problem, let us take a look at the example of Fig. 4.1 again. The original, complete histogram for this example is shown here in Fig. 4.2.

If we now divide the space of variable values into three intervals or subsets, namely 1:{1,2,3}, 2:{4,5,6}, and 3:{7,8}, we obtain the histogram of Fig. 4.3.

Then, if we split E_k into four subsets, 1:{1,2}, 2:{3,4}, 3:{5,6}, 4:{7,8}, we obtain the histogram of Fig. 4.4.

Another example of a four-segment histogram, shown in Fig. 4.5, is formed out of segments 1:{1,2,3}, 2:{4}, 3:{5,6}, 4:{7,8}.

The image from Fig. 4.4 provides, intuitively, a far better synthesis of data from Fig. 4.2 than that of Fig. 4.3. The reason is evident: in the case of Fig. 4.3, the interval (subset) 2 aggregates variable values, for which the numbers of objects are too different, and this difference disappears from the image. Similarly with Fig. 4.5,

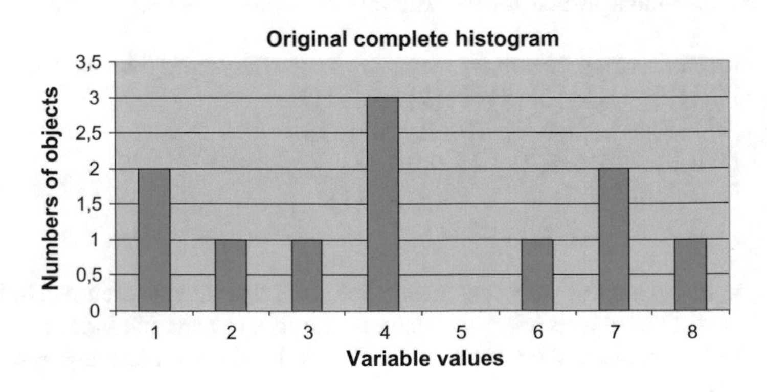

Fig. 4.2 Complete histogram of variable values from Fig. 4.1

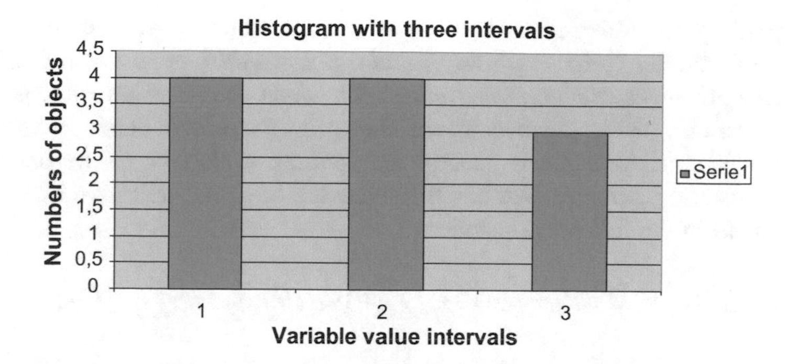

Fig. 4.3 Histogram for the case of Fig. 4.1 with three intervals

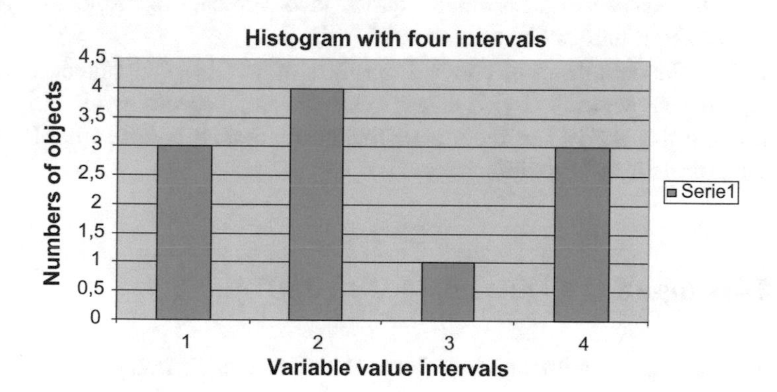

Fig. 4.4 Histogram for the case of Fig. 4.1 with four intervals

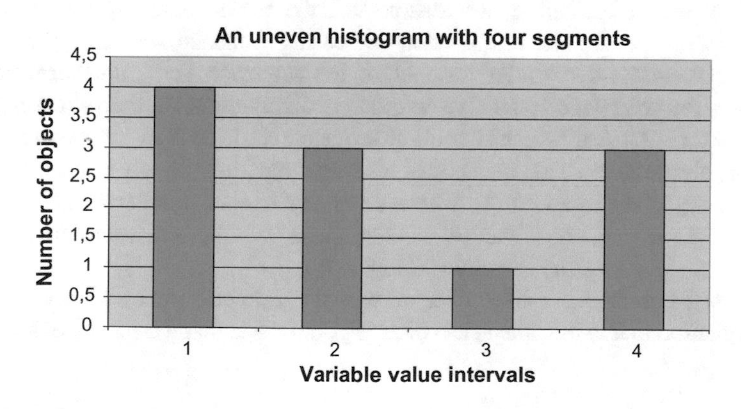

Fig. 4.5 Another four-segment-based histogram for the case of Fig. 4.1

corresponding to histogram, which separates the value of 4, with the highest number of objects. *Quite intuitively, again, one would like to get the intervals (subsets) that aggregate variable values, for which the numbers of objects are possibly similar, while separating the intervals, for which these numbers are possibly different.* And this is exactly the wording analogous to the one of the generic clustering problem. We can formulate the bi-partial objective function for this problem in the following manner:

$$Q_S^D(P) = C^D(P) + C_S(P) = \sum_c D(C_{ck}, C_{c+1,k}) + \sum_c S(C_{ck}), \qquad (4.4)$$

with the definitions following those of (4.1), but the transformation $s(d)$ taking the form $s(d) = (nd^{\max} - d)/2$, where $d^{\max} = \max_i d_{i,i+1}$. The division by two is introduced, because in this essentially quite "inconvenient" example[1] the sum of $d_{i,i+1}$ is only about half of the sum of $d^{\max} - d_{i,i+1}$.

Now, with the definitions as above, the values of the objective function (4.4) for the histograms shown in Figs. 4.2, 4.3, 4.4 and 4.5, are, respectively: 9, 7.5, 7.5 and 10.5, meaning that the last of them is distinctly the best. It is left to the Reader to check the rationality of this indication.

4.3 Division of the Univariate Distribution

We are here dealing with another form of the essentially very same problem, having, however, a different interpretation. Assume a univariate empirical distribution of a quantity x with values from $\boldsymbol{R}_+$. The distribution consists of n observations, indexed i, $i = 1, \ldots, n$. Denote the set of indices I. Without any loss to any sort of respective reasoning, we assume that the values taken by x in this distribution, denoted x_i, are ordered in a non-decreasing sequence, i.e. $x_{i+1} \geq x_i$ for all i.

Next, assume we consider, instead of the sequence $\{x_i\}_I$, the corresponding sequence, formed by the respective cumulative distribution, i.e. the values z_i defined as $z_i = \Sigma_{j=1,\ldots,i} x_j$. So, we deal with a sequence $\{z_i\}_I$ that is increasing and, in addition, also convex. This means that a straight line, joining any two points of the sequence $\{z_i\}_I$, say z_i and $z_{i+\Delta I}$, where Δi is any integer number contained in the interval $[2, n - i]$, has values above those of the corresponding z_i, i.e. $z_{i+1}, \ldots, z_{i+\Delta i-1}$ (see the example of Fig. 4.6).

For such Lorenz-curve-like[2] data we would like to construct a piece-wise linear approximation that is in some sense (actually, the sense that we promote throughout

---

[1]One would expect a realistic empirical distribution to approximate some known forms, while here it is, apparently, "haphazard".

[2]Actually, the proper Lorenz curve, used to represent the distribution of wealth or income (x_i) in some population, introduced in 1905 by M. O. Lorenz, concerns the normalised values, so that it always starts from 0 and ends with 1.

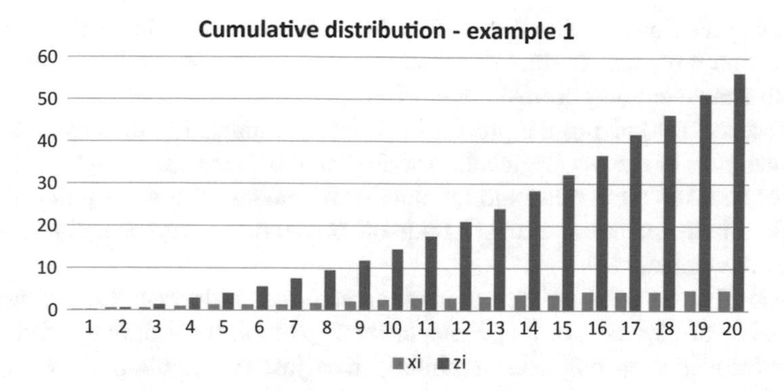

Fig. 4.6 An academic example of a convex cumulative distribution

this volume) "optimal". Namely, we would like to determine a set of line segments such that the resulting error (sum of absolute differences between the actual values of z_i and the corresponding values of the approximating function) is possibly low, while the number of segments distinguished is also kept reasonably low.

This problem, even if only roughly defined, applies not just to "approximation theory" (to which it actually does NOT apply), but to a wide variety of concrete domains. The one from which the particular motivation to undertake it came (see Owsiński, 2012b, c) was the distribution of social and/or economic indicator values among units indexed i (countries, regions, municipalities, etc.), for which we would like to obtain not "the best approximation", but the possibly "best" set of "classes" or "types", to which the units can be assigned. This is the case of some development indices, for which we seek an appropriate classification of, say, countries, into groups referred to as "highly developed", "developed", …, "dramatically lagging", without assuming an arbitrary division of index values i or thresholds in terms of x_i. Even more—we would not like to have the number of such classes defined beforehand, but, rather, obtained as the output from the procedure.

If we obtained such a "more objective" division, based only on the shape of the sequence $\{z_i\}$, then the assignment of labels, such as "highly developed" etc., would be done a posteriori on the basis of characteristics of the classes obtained, rather than, as this is very often done in practical terms, from a largely subjective perspective on how the classes "should" be defined or named.

The present analysis was motivated by exactly such a proposal from Nielsen (2011), concerning the country development levels. Another domain of interest with similar features is the one of distribution of wealth within a society, with the i's corresponding to some wealth classes, groups or even individuals. Nielsen's (2011) proposal is here analysed and extended on the basis of the "bi-partial" approach.

Like in the preceding cases, the exercise of distribution division is founded on the essential proposition that the respective distribution indeed displays the characteristics that make it susceptible for distinguishing discrete "levels", corresponding to different "classes" or "categories". Even if we admit that the convex cumulative distribution, like that of Fig. 4.6, conceals quite effectively in terms of visualisation any such distinct division, we suppose that it can be recovered in a justifiable manner using formal approaches.

It appears that solving the problem consists in minimising the error for subsequent numbers of classes (segments) and finding the "most appropriate" solution in terms of both the error value and the number of classes. The weak point of such a procedure is in finding a "proper" trade-off between the error magnitude and the number of segments.

Obviously, the error for the optimum approximation decreases as a function of the number of classes. So, it appears "natural" to look for a different form of the objective function to optimise (minimise), than just the (minimum) total error of approximation, namely, in very general terms, "total error + number of classes".

Conform to our notation, A_q will be the set of indices i of observations x_i (and so also z_i), classified in class q. We shall denote by $z^{q\max}$ and $z^{q\max}$, respectively, the minimum and maximum values of z_i, corresponding to the set A_q. These values, in turn, correspond to indices $i^{q\min}$ and $i^{q\max}$, respectively. Let us denote the set of i values, defining the partition of the sequence 1, …, n into the subsets A_q, i.e. the sequence composed of $1 = i^{1\min}, i^{1\max}, i^{2\min}, i^{2\max}, i^{3\min}, \ldots, i^{p\max} = n$, by ***iq***. By specifying ***iq***, we define $\{A_q\}$ and the entire solution. When referring to the explicit set of subsets $\{A_q\}$ we may also use the notation P, for partition of the set of observations.

For the assumed piece-wise linear approximation the general form of the qth piece is

$$z^q(i) = a^q i + b^q, \tag{4.5}$$

where we can no longer care whether i is discrete or continuous, as we observe the values only for natural i. The values of a^q and b^q are determined in a natural manner from the standard formulae, where we assume, formally, that each segment is composed of at least two consecutive observations, i.e. $i^{q\max} > i^{q\min}$:

$$a^q = \frac{z^{q\max} - z^{q\min}}{i^{q\max} - i^{q\min}}, \tag{4.6}$$

$$b^q = z^{q\min} - \frac{z^{q\max} - z^{q\min}}{i^{q\max} - i^{q\min}} i^{q\min}. \tag{4.7}$$

Note that after differentiating $z^q(i)$, given through the above formulae, we obtain the increasing sequence of levels a^q, corresponding to classes in terms of values of x_i.

In view of the convexity of the sequence of z_i, the sequence of a^q is non-decreasing, while the sequence of b^q is non-increasing.

We can now formulate the "minimum approximation error" problem, with the respective objective function, denoted $C_D(\{A_q\})$, as follows:

$$\min_{iq}\left(C_D(\{A_q\}) = \sum_q \sum_{i \in Aq} (z^q(i) - z_i)\right), \tag{4.8}$$

minimisation being performed with respect to the sequence ***iq***. We shall denote the optimum sequence, corresponding to the minimum in (4.8), by $\boldsymbol{iq}^*$.

As mentioned, the optimum value of this objective function is non-increasing in the number of segments, p (see the examples in Table 4.1, derived from the data shown in Fig. 4.6; although the technical assumptions differ among the examples, and explicit optimisation has not been carried out, the interpretation of the results appears to be obvious).

Since under convexity there is one optimum $\boldsymbol{iq}^*$ for each consecutive value of p (quite in line with Nielsen, 2011), we can denote the minimum value of $C_D(\{A_q\})$ for a given p by $C^*_D(p)$, so that $C^*_D(p) \geq C^*_D(p+1)$. Equality can only occur when sequences $x_i = x_{i+1} = \ldots$ exist, so that the corresponding values $z_i, z_{i+1}, \ldots$ lie on a straight line. Otherwise, any increase of p leads to a decrease of $C^*_D(p)$. One could go in this manner to the extreme of $p = n$, when $C^*_D(n) = 0$, an "ideal approximation"! Each observation would then constitute a separate "class" with one representative.

Obviously, when the already mentioned sequences $x_i = x_{i+1} = \ldots$, occur, so that the corresponding $z_i, z_{i+1}, \ldots$, are situated on a straight line, the value of $C^*_D(p)$, when p descends from n, shall remain at 0 also for several values of $p < n$, down to the value, determined by the total length of such uniform sequences.

While construction of approximating segments is not a question, the issue that we address here is related to finding a way to tell "how different the successive observations have to be in order to assign them to different segments (classes)?".

If the wording of this particular problem were "minimise the error with as low number of segments as possible", the following formulation would result:

$$\min\left(C_D(\{A_q\}) + w(p)\right) \tag{4.9}$$

where $w(p)$ is the weight attached to the number of segments. For consecutive values of p the minimum of $C_D(\{A_q\})$ would be found, and then the minimum of the function from (4.9) determined. Although this procedure might seem

Table 4.1 Examples of division of the cumulative distribution from Fig. 4.6 and the corresponding values of $C_D(\{A_q\})$

Number of segments	3	4	5	6
Subsets of indices forming the division	{1–3} {4–15} {16–20}	{1–5} {5–10} {10–15} {15–20}	{1–3} {4–7} {8–10} {11–15} {16–20}	{1–3} {4–7} {8–10} {11–13} {14–15} {16–20}
$C_D(\{A_q\})$	30.75	7.6	2.4	1.05

In this distribution $i = 1, \ldots, 20$

cumbersome, but, as we expect not too many segments to correspond to optimum, it would be still numerically quite feasible. We shall, though, not go into the technical details of such a procedure, for reasons given below.

Namely, now, the essence of the problem is transferred to determination of the function $w(.)$. In these cases, mostly of operational research or technical character, when p has a concrete interpretation, like, say, setup cost (e.g. of establishing and running p separate marketing schemes and actions for p categories of customers), while error minimisation leads to definite benefits (proper classification of a definite customer in terms of his/her profit for the company), then determination of $w(.)$ is feasible, even if difficult. This is not, however, the case with our problem, where we look for some possibly "natural" division of the distribution, and no cost/benefit, except for the facility of use of appropriate linguistic labels ("very highly developed", "highly developed",…), is involved.

As we try to find the "natural" division of the cumulative distribution (provided it exists, and the method we aim at ought to tell us somehow whether it does), therefore, we should refer to some "counterweight", analogous to that of $w(p)$ in (4.9), but having the same sort of meaning and kind of measurement as $C_D(\{A_q\})$, based on error. In this way we might be able to try to define the proper p and at the same time the ***iq***, or, otherwise, the $\{A_q\} = P$.

Thus, similarly as in (4.9), we would like to add to $C_D(\{A_q\})$ a component that would penalize, in this case, for the division into segments that are in some way "too similar", especially in terms of subsequent a^q. In general terms, the respective bi-partial objective function and the corresponding problem would look like

$$\min\left(C_D\left(\{A_q\}\right) + C^S\left(\{A_q\}\right)\right), \tag{4.10}$$

where $C^S(\{A_q\})$ corresponds to aggregate similarity between the consecutive segments, based primarily on differences of consecutive a^q. A concrete form of $C^S(\{A_q\})$ might be constructed as follows:

- first, a kind of difference between two consecutive segments, q-1 and q, is measured, from the point of view of the succeeding segment, q, as, for instance,

$$z_{iq\min} - a^{q-1} i^{q\min} - b^{q-1} \tag{4.11}$$

i.e. the difference between the actual value of z_i at the beginning of the next, qth subset of observations, and the "approximation" of the same, resulting from the previous segment. This difference is always non-negative, due to convexity of $\{z_i\}$, and can be interpreted as a "distance" between the two consecutive segments in the approximation;

- as we wish to penalize with the function $C^S(.)$ the *similarity*, not distance (difference), in order to convert (4.11) into similarity (proximity), we subtract it from an upper bound, which might, in particular, be constituted by the maximum of a similar difference for a given data set, namely the biggest difference of tangents along the curve of z_i, i.e. between its beginning and end; here, the two extreme tangents, $a^{(1)}$ and $a^{(n)}$, are defined as:

$$a^{(1)} = z_1/i_1;\ \text{and}\ a^{(n)} = (z_n - z_{n-1})/(i_n - i_{n-1}); \tag{4.12}$$

yet, in order to calculate the proper difference, we must have full expressions for the lines corresponding to $a^{(1)}$ and $a^{(n)}$, allowing for their use for consecutive subsets A_q; we assume, namely, that all four lines involved, corresponding to a^q, a^{q-1}, $a^{(1)}$ and $a^{(n)}$ cross at the point, defined otherwise by the crossing of the lines, corresponding to A_{q-1} and A_q (see the scheme of Fig. 4.7); from this condition we derive the values of b, to be used in conjunction with $a^{(1)}$ and $a^{(n)}$ (denoted, respectively, $b^{*(1)}$ and $b^{*(n)}$) in the appropriate expression, namely:

$$b^{*(1)} = b^q - (a^{(1)} - a^q)(b^{q-1} - b^q)/(a^q - a^{q-1}) \tag{4.13a}$$

$$b^{*(n)} = b^q - (a^{(n)} - a^q)(b^{q-1} - b^q)/(a^q - a^{q-1}). \tag{4.13b}$$

Now, the expression for $C^S(.)$ for a single q can be written down as

$$a^{(n)} i^{q\min} + b^{*(n)} - \left(a^{(1)} i^{q\min} + b^{*(1)}\right) - \left(z_{iq\min} - a^{q-1} i^{q\min} - b^{q-1}\right) \tag{4.14}$$

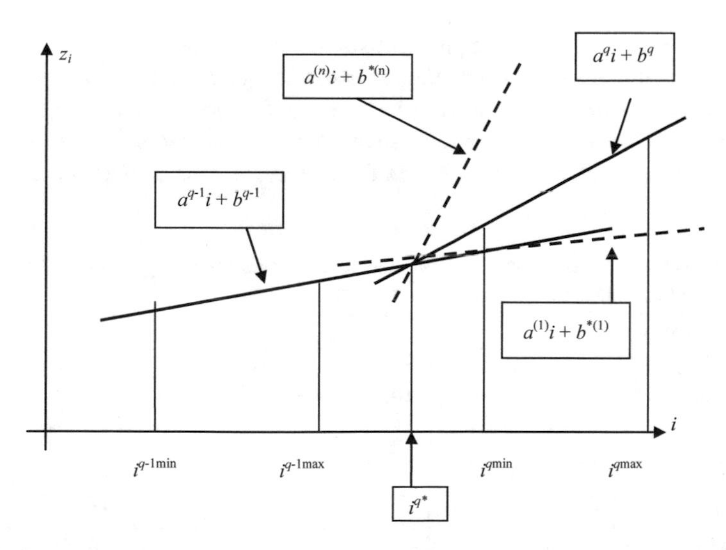

Fig. 4.7 The scheme of determination of similarities between the two consecutive approximating segments

where the second term in brackets is equivalent to the difference, given by (4.11), while the preceding terms define the reference for the given q. The proposed $C^S(P)$ is the sum over q of (4.14).

Altogether, the minimised objective function takes on the form:

$$\begin{aligned} C_D^S(\{A_q\}) &= C_D((\{A_q\}) + C^S(\{A_q\})) \\ &= \sum_q \sum_{i \in Aq} (a^q i + b^q - z_i) + \sum_q \left(a^{(n)} i^{q\min} + b^{*(n)}\right) \\ &\quad - \left(a^{(1)} i^{q\min} + b^{*(1)}\right) - \left(z_{iq\min} - a^{q-1} i^{q\min} - b^{q-1}\right), \end{aligned} \tag{4.15}$$

where we formally assume $a^0 = 0$ (which is natural) and $b^0 = 0$ (which is a bit artificial).

For the illustrative example considered here, the results for the divisions, already referred to in Table 4.1, taking, additionally, into account (4.15), are shown in Table 4.2.

The bi-partial objective function selects among the examples provided the one with five segments, its value for six segments being also higher than that for four. Since the respective partitions are (close to) nested, i.e. the increasing number of segments corresponds to divisions of selected A_q forming the preceding partition, this example shows that indeed we might deal with a convex objective function along such nested families of partitions. This implies the existence of a non-trivial minimum of $C_D^S(\{A_q\})$ in the set of all ***iq***, i.e. partitions, though we shall not be trying to demonstrate this here.

A Problem...

The above rationalization of the perspective on dividing the distributions of the kind we mean here, though, encounters often an essential, quite pragmatic, hindrance in the actual shapes of such distributions. An example, of quite a mild character at that, is given in Fig. 4.8, showing ordered (from the "best" to the "worst") country scores of the well known Quality of Life (QoL) ranking. See, for this and other instances, e.g.:

Table 4.2 Examples of division of the cumulative distribution from Fig. 4.6 and Table 4.1, and the corresponding values of $C_D(\{A_q\})$, $C^S(\{A_q\})$, and $C_D^S(\{A_q\})$

Number of segments	3	4	5	6
Subsets of indices forming the division	{1–3} {4–15} {16–20}	{1–5} {5–10} {10–15} {15–20}	{1–3} {4–7} {8–10} {11–15} {16–20}	{1–3} {4–7} {8–10} {11–13} {14–15} {16–20}
$C_D(\{A_q\})$	30.75	7.6	2.4	1.05
$C^S(\{A_q\})$	4.64	8.01	10.88	15.19
$C_D^S(\{A_q\})$	35.39	15.61	**13.28**	16.24

In this distribution $i = 1, \ldots, 20$

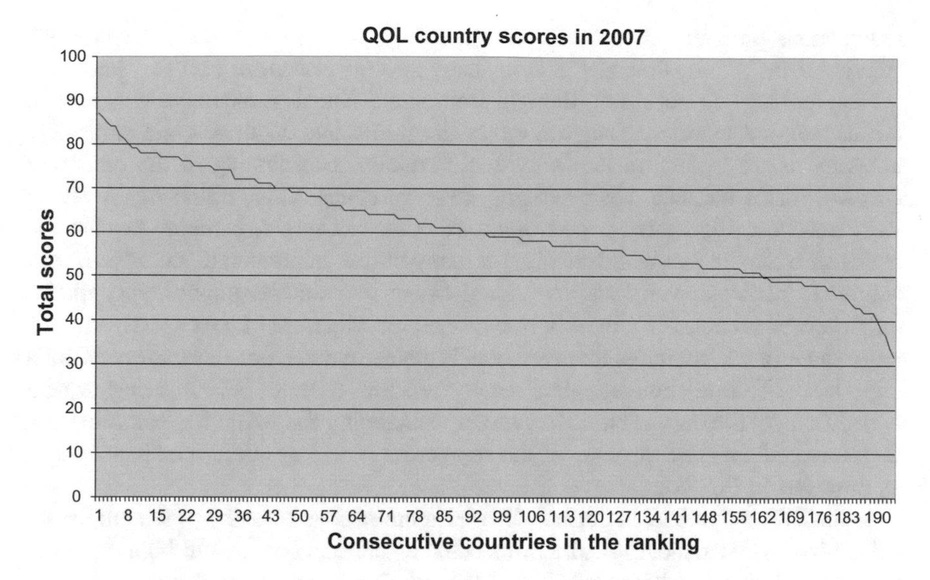

Fig. 4.8 An example of the total quality of life scoring, see http://www.il-ireland.com/il/qofl07/

http://www.economist.com/media/pdf/QUALITY_OF_LIFE.PDF—The Economist Intelligence Unit's quality-of-life index of 2005, as seen on Nov. 18th, 2011, or http://www.il-ireland.com/il/qofl07/—2007 Quality of Life Index), altogether for close to 200 countries of the world.

The total scores shown, ranging between 0 and 100, are based on nine partial scores for broad domains, such as "living costs", "economy", "environment", "freedom", "safety", etc. (for a deeper insight, consult Owsiński & Tarchalski, 2008).

The problem lies in the shape of the curve, corresponding to our sequence of x_i. One can easily see that—due to the nature of the scoring system—in the middle part of the curve there are numerous shorter and longer flat segments, some of them separated only by exactly minimum (the smallest possible) jumps. For this part of the curve the methodology here outlined, and also its broader rationality, can well be successfully applied. Yet, the two ends of the curve display a completely different character: sharp increase of the gradient towards the two extremes.

Within these two extreme parts of the curve the methodology—and the broader rationality—would have to distinguish several classes, with very few objects, indeed, in most cases—just one—in each consecutive class. This seems to bend the rationality we made use of. Also the approach of Nielsen (2011) will have troubles with this shape.

This shape, however, is not an incidental result of the methodology, adopted in creating the QoL ranking and the actual data used. It is a consistent feature—the very same shape appears in most of the partial score-based rankings. It is even much more pronounced in some of them (e.g. for "living costs", "economy", "environment", "infrastructure"). This shape appears also from year to year.

The rankings do not result, though, from some statistical measurement, at least not as they are reported. They are either the immediate result of quite subjective assessments of experts on the individual variables, contributing to the particular domains, or of the data, characterising these variables. Thus, ultimately, we deal with somehow aggregated expert opinions. This fact might, perhaps, explain the character of the final output. Namely, for many of the "intermediate" countries, with respect to particular variables, individual expert assessments might barely distinguish between them, while the extremes are, at least by most of the experts, and in a large share of the countries involved, easily noted. Now, since there is generally a high level of correlation between many variables (roughly +70% being typical correlation coefficient), such observations, concerning the extremes, summing up, therefore, and creating the ends of the respective curves, result in the "bent" edges as observed in Fig. 4.8.

Actually, the issue is, in general, insofar more serious as many of the empirical, "objective" or "statistical" distributions behave, indeed, according to highly regular functional shapes (see the case of Fig. 4.9, related to an entirely different case, when the measurements are fully "objective": the running times in a race), so that there is very little ground for dividing them in a different manner than on the basis of *substantive criteria* (e.g. the "biological minimum" or "social minimum" thresholds in the case of poverty). Application of the approach outlined here would then involve the measures of fit to/divergence from the matching functional shapes, which is a fully feasible option, in which the respective formulae would simply change their form, while the sense of the approach would remain the same.

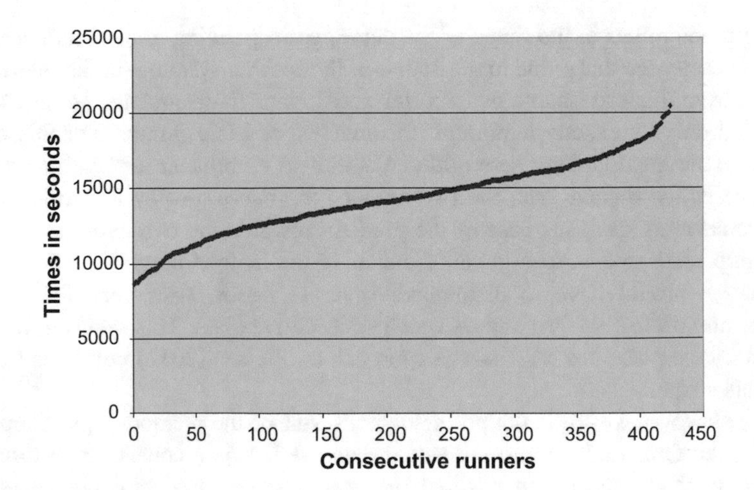

Fig. 4.9 Results (time in seconds) of the runners in the marathon race ("Koral marathon") in Krynica, Poland, in 2012 (diagram shows the data published in "Rzeczpospolita" daily)

Hence, the following question arises in the context of the optimum distribution division problem:

- if the results of a division exercise indicate a similar phenomenon to that here commented upon, indicating a sort of (unexpected?) regularity, can we deduce something about the way in which the respective distribution has been constructed?

Acknowledgements
The Research Reported in This Particular Subsection Was Partly Supported Within the Projects: TIROLS, Funded by the Polish Ministry of Science and Higher Education, N N516 195237, and "Development Trends of Masovia", MBPR/W/TRM-0712-08-2011, Co-Financed from the EU Funds.

4.4 The p-Median/p-Center Facility Location Problem

Problem Formulation

The problem we address here is different from the majority of problems taken as instances of application of the bi-partial approach. It is a classical question in operations research, related to location analysis. Not only, though, the interpretation of the problem is quite specific, but also the very form is in a way not appropriate for the treatment through the bi-partial formalism, as introduced here.

We deal, namely, in a very simplistic, but also very general manner, with the following problem

$$\min \Sigma_q \left(\Sigma_{i \in A_q} d(x_i, x^q) + c(q) \right) \tag{4.16}$$

with minimisation being performed over the choice of the set of p points (objects) x_i that are selected as the central or median points $x^q, q = 1, \ldots, p$. For our further considerations it is of no importance, whether these points belong to the set X of objects (medians) or not—i.e. they are only required to be the elements of the space E_X (centers). It is, however, highly important that the second component of the objective function, namely $\Sigma_q c(q)$, does not involve any notion of distance or proximity.

While $d(.,.)$ is some distance, like in all problems here considered, $c(q)$ is a non-negative value, interpreted as cost, related to a facility q. The problem, namely, is to find a set of p ($q = 1,\ldots,p$) locations of facilities, such that the overall cost, composed of the sum of distances between points, assigned to the individual facilities, and these facilities, and the sum of costs, related to these facilities, is minimised. It is, of course, assumed that the costs $c(q)$ and distances $d(.,.)$ are appropriately scaled (or measured), in order for the whole to preserve interpretative sense.

The costs $c(q)$ may be given in a variety of manners: as equal constants for each arbitrary point from X or from E_X, i.e. c, so that the cost component in (4.16) is simply equal pc, or as (more realistically) the values, determined for each point separately, i.e. $c(i)$, or as a function, composed of the setup component (say, c_1, if this setup cost is equal for all locations) and the component that is proportional to the number of locations, assigned to the facility q, with the proportionality coefficient equal c_2 (i.e. the cost for a facility is then c_1 + card$A_q c_2$). Of course, more complex, nonlinear cost functions, also with c_1 replaced by $c_1(i)$, can, as well, be (and sometimes are) considered.

This problem has a very rich literature, with special numerical interest in its "pure" form, without the cost component, mainly devoted to mathematical and geometric properties and the respective (approximation) algorithms and their effectiveness. Notwithstanding this abundant tradition, the issues raised and the results obtained, we shall consider here the form of (4.16) in one of its basic variants.

Some Hints at Cluster Analysis

Any Reader with some basic knowledge in cluster analysis shall immediately recognise the first component of (4.16) as corresponding to the vast family of the so-called "k-means" algorithms, where such a form is taken as the minimised objective function. Indeed, this fact is the source of numerous studies, linking facility location problems with clustering approaches. One can cite in this context, for instance, the work of Pierre Hansen (e.g. Hansen, Brimberg, Urosević, & Mladenović, 2009), but most to the point here is the relatively still recent proposal from Liao and Guo (2008), which explicitly links k-means with facility location, similarly as this was done several decades ago by Mulvey and Beck (1984).

The latter proposal, formulated by Liao and Guo (2008) is insofar interesting as the ease of realisation of the basic k-means algorithm (to which we shall yet return in Sect. 5.2 of the present book) allows for the relatively straightforward accommodation of additional features of the facility location problem (e.g. definite constraints on facilities and their sets).

An Example

For the sake of illustration, we shall treat the problem (4.16) in the following more concrete form:

$$\min \Sigma_q (\Sigma_{i \in A_q} d(x_i, x^q) + c_1 + c_2 \mathrm{card} A_q) \tag{4.17}$$

where c_1 is the "facility setup cost", while c_2 is the unit cost, associated with the servicing of each object $i \in A_q$. This formulation, even if still quite stylised, seems to be plausible as an approximation.

The form (4.17) can be, quite formally, and with all the obvious reservations, which were mentioned, anyway, before, moulded into the general bi-partial scheme, i.e.

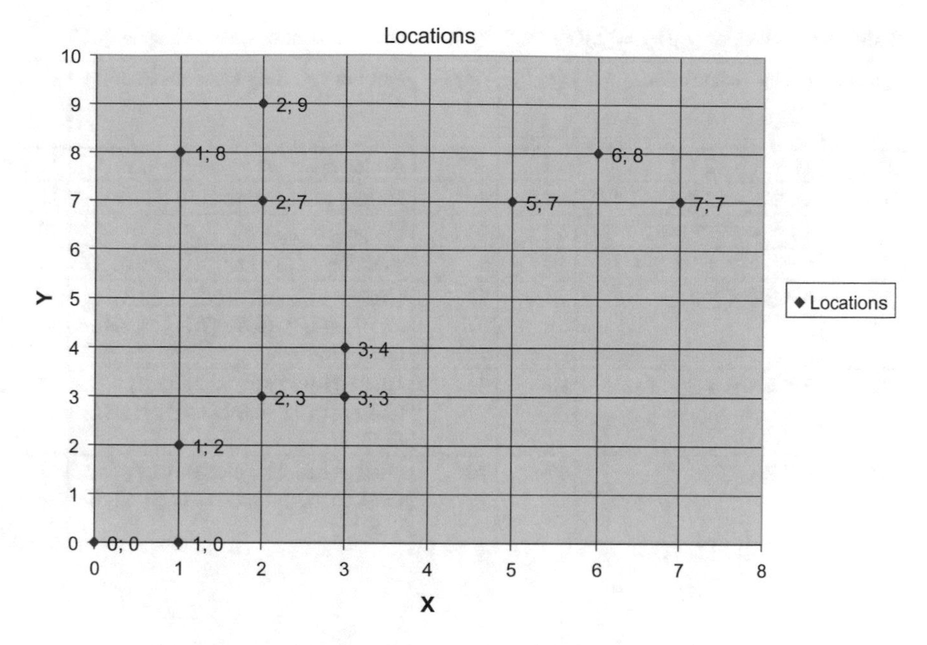

Fig. 4.10 A simple example for the facility location problem

$$\min_P Q_D^S(P) = Q_D(P) + Q^S(P), \tag{4.18}$$

where partition P encompasses, in this case, both the composition of A_q, $q = 1,\ldots,p$, taken together with the number p of facilities, and the location of these facilities, i.e. choice of locations from X as the places for facilities q.

Consider a simple case, which is shown in Fig. 4.10, with the distance $d(.,.)$ defined as Manhattan distance, and the cost component of (4.17) being based on the parameter values $c_1 = 3$, $c_2 = 1$. Again, these numbers, if appropriately interpreted, can be considered quite plausible (for instance, distance, corresponding to annual transport cost, and c_1 corresponding to annual write-off value).

Table 4.3 shows the exemplary values of $Q_D^S(P) = Q_D(P) + Q^S(P)$, according to (4.17), for a series of partitions P. This is a nested set of partitions, i.e. in each consecutive partition in the series one of the subsets of objects, cluster A_q is the sum of some of the clusters from the preceding partition, with all the other clusters being preserved. Such a nested sequence of partitions is characteristic for a very broad family of cluster algorithms—the progressive merger or progressive split algorithms.

The character of results from Table 4.3, even if close to trivial, is quite telling, and indeed constitutes a repetition of the observations made for other cases here considered. Note that the values of $Q_D(P)$ increase along the series of partitions, while the values of $Q^S(S)$—decrease, and $Q_D^S(P)$ has a minimum, which, for this simple case, corresponds, indeed, to the solution to the problem.

Table 4.3 Values of $Q_D^S(P) = Q_D(P) + Q^S(P)$ for a series of partitions, according to (4.17)

$Q_D(P)$	$Q^S(P)$—calculation	$Q^S(P)$—value	$Q_D^S(P)$	Partitions (facility locations in bold)	p
0	12 * 3 + 12 * 1	48	48	All locations are facility locations	12
1	11 * 3 + 10 * 1 + 1 * 2	45	46	Merger of (0,0) and (1,0)	11
2	10 * 3 + 8 * 1 + 2 * 2	42	44	Merger of (2,3) and (3,3)	10
3	9 * 3 + 7 * 1 + 2 + 3	39	42	Addition of (3,4) to (2,3) and (3,3)	9
13	4 * 3 + 4 * 3	24	37	{(0,0), **(1,0)**, (1,2)} {(2,3), **(3,3)**, (3,4)}, {(5,7), **(6,8)**, (7,7)}, {**(1,8)**, (2,7), (2,9)}	4
22	3 * 3 + 6 + 3 + 3	21	43	{(0,0), (1,0), **(1,2)**, (2,3), (3,3), (3,4)}, {(5,7), **(6,8)**, (7,7)}, {**(1,8)**, (2,7), (2,9)}	3
55	1 * 3 + 12	15	70	{(0,0), (1,0), (1,2), (2,3), (3,3), **(3,4)**, (5,7), (6,8), (7,7), (1,8), (2,7), (2,9)}	1

Some Algorithmic Considerations

As indicated before, the problem lends itself to the k-means-like procedure, which, in the general and quite rough terms, at that, takes the following course:

0° Generate p[3] points as initial (facility location) seeds (in this case the points generated belong to X), usually $p \ll n$
1° Assign to the facility location points all the n points from the set X, based on minimum distance, establishing thereby clusters A_q, $q = 1,\ldots, p$
2° If the stop condition is not fulfilled, determine the representatives (facility locations) for the clusters A_q, otherwise STOP
3° Go to 1°.

This procedure, as we know, converges very quickly, although it can get stuck in a local minimum. Yet, owing to its positive numerical features, it can be restarted from various initial sets of p points many times over, and the minimum values of the objective function obtained indicate the proper solution.

In the case here analysed, since the facility location problems rarely are really large in the standard sense of data analysis problems, it is quite feasible to run the k-means procedure, as outlined above, for consecutive values of p in order to check whether (for what value of p) a minimum over p can be found for a definite formulation of the facility-location-related bi-partial objective function $Q_D^S(P)$.

[3]We use the classical name of the k-means algorithm, although throughout this book the number of clusters, referred to in this name as "k", is denoted p.

4.5 Block-Diagonalisation or Concept Identification

Problem Statement

Consider the binary data matrix like, for instance, that of Fig. 4.11. Assume that objects (rows) are machines, while variables (columns) represent parts or operations, potentially executed with the machines represented by the rows, and the entire matrix represents a technological process, or a set of closely linked technological processes. This kind of matrix is called "incidence matrix", denoting the fact that only some variables are "incident" with some of the definite objects.

We would like to rearrange this matrix, by permuting its rows and columns appropriately, so as to form the blocks of 1's, meaning groups of machines, performing similar operations, or operating on similar sets of parts. This is called "cell formation" in the domain of "flexible manufacturing" (see, e.g., Owsiński, 2001b, 2009 as well as Owsiński, Stańczak, Sęp, & Potrzebowski, 2010). The more coherent groups we are able to form (see Fig. 4.12), i.e. the less 1's there are outside blocks and 0's inside blocks, the less we lose on transport, regime change, material flow and operation cost in production, carried out on the basis of the thus designed manufacturing cells. Altogether, we wish to obtain, for the given data set, the possibly "clean" cells for production,

Objects i / variables k	k=1	2	3	4	5	6	7
i = 1	1	0	1	0	0	1	1
2	0	0	0	0	1	0	1
3	0	0	1	1	1	0	0
4	1	0	1	0	0	0	0
5	0	1	0	0	0	1	1
6	1	1	0	0	0	1	0
7	1	0	1	1	1	0	0
8	1	0	1	1	0	1	1

Fig. 4.11 A simple example of an incidence matrix

Objects i / variables k	k=1	6	3	4	5	2	7
i=1	**1**	**1**	**1**	0	0	0	1
4	**1**	**0**	**1**	0	0	0	0
6	**1**	**1**	**0**	0	0	1	0
8	**1**	**1**	**1**	1	0	0	1
7	1	0	1	**1**	**1**	0	0
3	0	0	1	**1**	**1**	0	0
2	0	0	0	0	1	**0**	**1**
5	0	1	0	0	0	**1**	**1**

Fig. 4.12 Incidence matrix from Fig. 4.10 after a permutation of rows and columns aiming at block-diagonalisation

This problem is quite different from the two previous ones in that, *first*, we deal with multiple dimensions ($k = 1, \ldots, m$) at the same time, *second*, we appear to be grouping, simultaneously, *objects and variables* so as to form blocks, and *third*, the blocks (groups) *need not be disjoint*. Thereby, while in most of the preceding cases we dealt, ultimately, with just a simplified, even though specific, clustering problem, we deal here with a problem that seems to exceed the basic formulation of the clustering problem[4].

Yet, despite these differences, the fundamental paradigm holds: ***similar machines (objects) ought to be grouped together, while dissimilar ones ought to be put separately***, similarity (proximity) meaning 1's and 0's in possibly the same columns (variables). The actual ("visual") formation of blocks can, namely, be considered a task apart, even though its actual complexity turns out to be comparable with that of the "mere" grouping of objects.

Concept Identification

The problem that was illustrated here with "machines" vs. "parts", or "operations", has a much more general interpretation. The blocks can, namely, be interpreted as "*concepts*" :sets of objects, characterised by (possession of) the same sets of features (1 for having a feature and 0 for not having it), see Bock (1994). We have already alluded to this kind of (language-oriented) interpretation, when commenting upon Fig. 2.1. The sets of objects, corresponding to clusters (better: not just those objects contained in the set X, but all of those potentially classified in the region of the space E_X corresponding to the given clusters) are called *extensions* of the concepts, while the sets of variables in the block—*intensions*. Naturally, concepts need not be mutually disjoint in terms of variables (intension), although they tend to be rather disjoint in terms of objects (extension), this being clearly closely connected with the language use of concepts.

One more remark is due at this point. In both these interpretations of the block-diagonalisation problem, we have been referring to the binary incidence matrix. Actually, in either case we may have, as data, integer or real numbers, expressing some sort of "intensity" rather than only "incidence". This may correspond, in the "machines-parts" interpretation, to cost or time of operations, and in the (language-wise) "concept" interpretation—to the intensity or magnitude of a given feature for a given object (just as shown in Fig. 2.1). Interestingly, this does not change in any way the formulation, related to cluster analysis, and, in particular, the formulation of the bi-partial objective function, nor does it change the approach as such.

We may use in this case the concrete objective function, defined below in (3.3), i.e.:

[4]It is obvious that with this problem we definitely go back to the basic insights of Jan Czekanowski's from the beginnings of the 20th century, who studied the serial resemblance of the skulls of Neanderthals (see Czekanowski, 1909, 1913, 1926, 1932).

$$
\begin{aligned}
Q_S^D(P) &= C_S(P) + C^D(P) = \sum_q S(A_q) + \sum_{q' > q} D(A_{q'}, A_q) \\
&= \sum_q \sum_{i<j \in Aq} s_{ij} + \sum_q \sum_{q' > q} \sum_{i \in Aq} \sum_{j \in Aq'} d_{ij}
\end{aligned} \tag{3.3}
$$

with an appropriate definition of s_{ij}.

By taking the above formulation, we explicitly neglect the fact that clusters (blocks) may effectively overlap, thus relegating the task of actual delineation of clusters with their overlappings to the hypothetical post-processing phase.

We could, however, also formulate an entirely analogous objective function for the case with explicitly overlapping clusters:

$$
\begin{aligned}
Q_S^D(P) &= C_S(P) + C^D(P) = \Sigma_q S(A_q) + \Sigma_{q' > q} D(A_{q'}, A_q) \\
&= \sum_q \sum_{i<j \in Aq} s_{ij} + \sum_q \sum_{q' > q} \sum_{i \in Aq' \setminus Aq \cap q'} \sum_{j \in Aq' \setminus Aq \cap Aq'} d_{ij}.
\end{aligned} \tag{4.19}
$$

In (4.19) we omit in the expression for $D(A_q, A_q)$ the common part $A_q \cap A_{q'}$, as having already been accounted for in both $S(A_q)$ and $S(A_{q'})$.

For the example, shown in Figs. 4.11, 4.12 and 4.13, we can calculate the values of the objective functions provided above, i.e. (3.3) and (4.19), for different definitions of the transformation $s(d)$. The respective values are provided in Fig. 4.14.

Thus, according to the objective function (3.3), it is the second of configurations from Fig. 4.14 (corresponding to Fig. 4.12) that is the best of those there considered, while according to the objective function (4.19) it is the second or the fourth one, depending upon the proximity definition applied.

Actually, as one looks at Figs. 4.11, 4.12 and 4.13, the two configurations appear to be at least plausible, if not "the" best. The hesitation is a counterpart to the existing multiplicity of criteria, that may be applied to this problem, and the different distance definitions that can be used, meant to best represent the practical side of the problem, as surveyed, for instance, in Owsiński (2009).

Objects *i* / variables *k*	*k*=1	6	3	4	5	2	7
i = 1	**1**	**1**	**1**	0	0	0	1
4	**1**	**0**	**1**	0	0	0	0
6	**1**	**1**	**0**	0	0	1	0
8	**1**	**1**	**1**	1	0	0	1
7	***1***	***0***	***1***	1	1	0	0
3	0	0	1	**1**	**1**	0	0
2	0	0	0	***0***	***1***	0	1
5	0	1	0	0	0	**1**	**1**

Fig. 4.13 Another view at the blocks from Fig. 4.12

Partitions into blocks	Objective function (3.3)		Objective function (4.19)	
	$s =$			
	$d^{max} - d$	$d^{max} + d^{min} - d$	$d^{max} - d$	$d^{max} + d^{min} - d$
{1}{2}{3}{4}{5}{6}{7}{8}*	104	104	104	104
{1,4,6,8}{3,7}{2,5}*	120	128	120	128
{1,4,6,7,8}{2,3}{5}*	115	126	115	126
{1,4,6,***7***,8}{2,3,***7***}{5}	138**	151**	119	132
{1,4,***6***,***7***,8}{***2***,3,***6***,***7***}{***2***,5}	156**	172**	112	128
{1,2,3,4,5,6,7,8}*	92	120	92	120

7 – figures in italicised bold denote overlappings; * values of objective functions (3.3) and (4.19) for these partitions are the same, as there are no overlappings; ** the objective function (3.3) is actually inapplicable to these cases, as it double counts the overlappings

Fig. 4.14 Various evaluations of partitions into blocks for the example of Figs. 4.11, 4.12 and 4.13

4.6 Rule Extraction

Problem Outline

Rule extraction is a very important problem in data analysis. It has received a great deal of attention over the last couple of decades (see, e.g., Piatetsky-Shapiro, 1991; Agrawal, Imielinski, & Swami, 1993; Hilderman & Hamilton, 2001; Bramer, 2007; Greco, Słowiński, & Szczęch, 2009). Rule extraction, or identification, is of special importance for the control- or policy-oriented applications, and, in a particular manner, for management. It takes a very special place in medical applications, both on the stage of diagnosis and in determining appropriate medication.

The problem, as we consider it here, this particular formulation not necessarily following the most frequent encountered ones, consists in determining, on the basis of the data matrix X, given the division of the variables into two groups, K^1 and K^2, $K^1 \cup K^2 = K$, one of them containing "input (explanatory) variables" and the second—"output (explained) variables", an *optimum set of rules*, defining the conditions for obtaining all the (essential) values of the "output variables"[5], in the general form

[5]Although the rules extracted take usually the form of "if… then…" clauses, the reference to causal implication is often explicitly avoided. The expressions, appearing in the rules are referred to, alternatively, respectively as: "antecedent", "premise" or "condition", and "consequent", "conclusion", "decision" or "hypothesis".

$$\text{IF}(x_{.k1} \in E(k1,1)) \wedge (x_{.k2} \in E(k2,1) \wedge \ldots$$
$$\text{THEN}(x_{.l1} \in E(l1,1)) \wedge (x_{.l1} \in E(l2,1) \wedge \ldots$$
$$\text{IF}((x_{.k1} \in E(k1,2)) \wedge (x_{.k2} \in E(k2,2) \wedge \ldots$$
$$\text{THEN}(x_{.l1} \in E(l1,2)) \wedge (x_{.l2} \in E(l2,2) \wedge \ldots$$
$$\text{IF}(x_{.k1} \in E(k1,r*)) \wedge (x_{.k2} \in E(k2,r*) \wedge \ldots$$
$$\text{THEN}(x_{.l1} \in E(l1,r*)) \wedge (x_{.l2} \in E(l2,r*) \wedge \ldots$$

where $k1$, $k2$,... denote the indices of "input variables", and $l1$, $l2$,... - the indices of "output variables"; $E(k,r) \in E_k$ are the sets of values, taken in the rules numbered r, by the variables k. It is also assumed that $E(lt,r), r = 1,2,\ldots$, appearing in the set of rules, exhaust all the feasible values of $x_{.lt}$ for all t, i.e. for all the output variables. This means that the rules define the input conditions, under which all the feasible output values are obtained.

The "optimum" mentioned with respect to the set of rules that one wants to extract from a given set of data X refers to two opposing aspects of the problem:

- the rules should be possibly simple (short), i.e. they ought to involve as few variables on the "input" side as possible, and the sets $E(.,.)$ ought to be possibly simply expressed (most desirably: a single number, or a single interval limit, like $> a$);
- at the same time, the rules ought to be possibly precise, in the sense that there should be in the data set X a possibly low number of exceptions to the rule system, i.e. the cases, when (i) rule conditions, or antecedents, are satisfied, but the output variables, or the consequents, do not take the prescribed values, and (ii) the rule conditions are not satisfied, but the output variables take the values envisaged in the rule system.

Sometimes this opposition, perceived, though, in a somewhat different perspective, is also being referred to through the notions of "fit" and "generalisation". Since the choice of the (set of) rules is not uniquely defined, the studies in rule extraction devote a large share of attention to the (development of the) criteria used and to their properties (see, e.g. Greco, Słowiński, & Szczęch, 2009).

So, an ideal system of rules would consist of few single-variable rules, depending upon single values of these variables, or single intervals of values, and would be absolutely precise, i.e. without any exceptions, whether of type (i) or (ii), as defined above (the *perfect rules*).[6]

It is possible, for many data sets, to construct the perfect rules, provided there are no contradictory observations (i.e. the ones with the same antecedents, but different consequents). This, however, usually would mean very long and elaborate rules, and/or a very high number of them, far from the satisfaction of the first aspect of the

[6]It should be noted that simplicity of rules is closely associated with their "intuitive appeal" and with practicability (effectiveness) of use in many circumstances (e.g. when they are used or at least consulted by human operators).

problem (due to the low degree of generalisation and high sensitivity to new observations).

Let us repeat the reservation that the formulation of the rule identification problem that is provided here is by no means generally admitted. Thus, on the one hand, it is not "sufficiently general" as it does not account for relations among the "input variables", for instance, and, on the other hand, in most of the actually treated cases, either theoretically or practically, it is very importantly simplified. The most significant, but straightforward, simplification is the separate treatment of [each of] the "output variables"; another lies in the binary (or just few) values of all variables.

The above reservation is insofar important as in the general view the rules can be regarded as an expression of functions, yielding the values of the "output variables" for given values of the "input variables". The form of rules rather than any other (e.g. algebraic expression) is simply more convenient exactly in cases when the respective spaces of variable values, E_k and E_l, are highly discrete, i.e. contain relatively few values, either by their very nature, or through categorisation.

An Example

For the sake of illustration, Table 4.4 contains a simple example for a set of binary variables. In this table, conform to the notation, the "input variables" are denoted with index k, and the "output variables" with index l.

The rules that fit perfectly the data from Table 4.4 are

$$\begin{array}{ll} \text{IF}(x_{k1}+x_{k2}+x_{k3}=2) & \text{THEN}\, x_{l1}=1 \\ \text{IF}(x_{k4}=1 \wedge x_{k1}x_{k2}x_{k3}>0) & \text{THEN}\, x_{l2}=1 \end{array}$$

meaning that otherwise the two "output variables" are equal zero. These perfect rules, i.e. without any exception, do not conform to the general form specified before, as we introduced two different functions of several variables, which, of course, is of no more general importance for the issue. The fact that these rules are perfect and the data are binary allows also for writing them down as straight logical expressions.

We have used a binary example for illustration not without cause. It is not only simple, but, for technical reasons, is actually the most often treated case, since other cases require, almost by definition, much more complex approaches and yield often too cumbersome results for practical use. The simplicity makes it possible to analyse the essential aspects of the problem by visual inspection.

We shall now illustrate this capacity by transforming the matrix of observations from Table 4.4. Table 4.5 a and b show the same entries with (a) separation of sub-matrices for the different combinations of the "output variables", and (b) a slight transformation, consisting in reordering the rows of the matrix, this transformation meant to show a bit more clearly the potential patterns of data, corresponding to rules.

The image of Table 4.5 b provides exactly the impression that we aim at, namely that a definite similarity of rows can be observed, and, indeed, taken advantage of, as facilitating the determination of the rules we look for.

Table 4.4 An example for the rule extraction problem

Variables: objects	$k = 1$	2	3	4	5	$l = 1$	2
$i = 1$	0	0	0	1	0	*0*	*0*
2	1	0	1	1	1	*1*	*1*
3	1	1	0	0	0	*1*	*0*
4	1	1	1	0	1	*0*	*0*
5	0	1	0	0	0	*0*	*0*
6	0	1	1	1	0	*1*	*1*
7	0	0	0	1	1	*0*	*0*
8	1	1	1	0	0	*0*	*0*
9	0	0	1	0	1	*0*	*0*
10	0	1	1	0	0	*1*	*0*

Table 4.5 An example for the rule extraction problem: separation of observations for combinations of "output variable" values

(a)

Variables k/l	$k = 1$	2	3	4	5	$l = 1$	2
Objects i: $i = 1$	0	0	0	1	0	*0*	*0*
4	1	1	1	0	1	*0*	*0*
5	0	1	0	0	0	*0*	*0*
7	0	0	0	1	1	*0*	*0*
8	1	1	1	0	0	*0*	*0*
9	0	0	1	0	1	*0*	*0*
3	1	1	0	0	0	*1*	*0*
10	0	1	1	0	0	*1*	*0*
2	1	0	1	1	1	*1*	*1*
6	0	1	1	1	0	*1*	*1*

(b)

Variables: objects	$k = 1$	2	3	4	5	$l = 1$	2
$i = 1$	0	0	0	1	0	*0*	*0*
7	0	0	0	1	1	*0*	*0*
9	0	0	1	0	1	*0*	*0*
5	0	1	0	0	0	*0*	*0*
4	1	1	1	0	1	*0*	*0*
8	1	1	1	0	0	*0*	*0*
3	1	1	0	0	0	*1*	*0*
10	0	1	1	0	0	*1*	*0*
6	0	1	1	1	0	*1*	*1*
2	1	0	1	1	1	*1*	*1*

Hence, without going into further details—evidently, the "philosophy" of the bi-partial approach can be used in solving the rule extraction problem, with the partition of the set of objects into subgroups (clusters) corresponding to the individual rules. Actually, the tradeoff between "fit" and "generality" or "precision" and "simplicity" forms a striking counterpart to the bi-partial approach. In order to apply this approach, the respective objective function ought to be devised, following the rules described here already several times over.

4.7 A More General Analogy

The problem considered in the preceding section is, in general, "statistical" in that we do not really hope to obtain the perfect rules, just as we did not hope to obtain perfect block-diagonalisation (nor do we expect to obtain "perfect clusters"). Actually, there is a high degree of analogy between the two kinds of problems, which goes well beyond the proximity (similarity) of objects. Let us recall that in the block-diagonalisation we tried to minimise the number of exceptions, i.e. 0's inside the blocks of 1's, and, at the same time, vice versa, the number of the opposite exceptions—1's outside blocks. In the case of rules we would also like to minimise the number of exceptions—observations, where antecedents ought to produce a rule-defined consequent, but do not, and, simultaneously, vice versa—observations, where antecedents ought not, according to the rules, produce a given consequent, but do so.

The issue in both cases is that (a) we cannot hope for perfect structures (blocks, rules), in view of "imperfect" data, and (b) even if we could have the perfect structures, they might consist of too many elements (blocks, rules)—ultimately, in order to avoid any imperfections, of n rules!—to be of any practical use.

The above description is also very much remindful of the problem of classification, in conformity with the very rough scheme shown in Fig. 4.15.

				Actual or "ideal" situation	
	Block-diagonalisation			1 in a block	1 outside of a block
		Rule identification		Observation follows rule q	Observation follows another rule
			Classification	Object in class q	Object in another class
Output from an algorithm	1 in a block	Antecedent follows rule q	Object classified in q	**a**	**b**
	1 outside of a block	Consequent does not follow rule q	Object classified otherwise	**c**	**d**

Fig. 4.15 A scheme of analogous interpretations related to basic criteria in classification, block-diagonalisation and rule extraction; **a**, **b**, **c** and **d** denote numbers of respective objects

This scheme omits certain situations, and does not account for all the actually made distinctions, because it serves the purpose of showing just the analogy mentioned. This is of special significance for rule identification and classification problems, whose analogy is straightforward. In fact, rules can generally be treated as classification rules in that they assign a set of input combinations to an output label, a class.

The numbers of respective objects or the elements considered, which are shown in Fig. 4.15 as denoted **a**, **b**, **c** and **d**, are usually used to form the criteria of goodness of the algorithms applied. Definitely, we try to make especially **a** (but also **d**) possibly high for a given data set, while trying to minimise **c** and **d** (a Reader is referred again to Owsiński, 2009, for a partial survey).

The bi-partial approach that we introduce here is exactly meant to provide a natural balance to all kinds of similar situations, without the necessity of making explicit reference to often cumbersome details of objective functions built on the basis of distinctions here shown.

There is also another essential link between the respective situations, namely that of definition of distances or proximities, and their close relation with the issue of the choice of criteria. We mentioned already that in the case of block-diagonalisation the choice of distance definition was quite broad, driven by the respective technical problem formulation, and in the case of rules the division into "input" and "output" variables poses an additional difficulty.

The issue of the definition of distance is in the case of rules indeed very closely associated with, if not contained in, the already indicated multiplicity of criteria and their choice. Without going into details of respective analysis we can simply propose that for this case distances (or proximities) be defined via a parametric definition:

$$d_{ij} = w^K d_{ij}^K + \left(1 - w^K\right) d_{ij}^L, \tag{4.20}$$

where, conform to the notations adopted, d_{ij} is the distance between observations (rows) i and j, d_{ij}^K is the distance between i and j, calculated over the set of variables indexed k ("input variables"), d_{ij}^L is the distance for the set of variables indexed l (one or more "output variables"), and $w^K \in [0, 1]$ is the weight, assigned the entire set of "input variables". The option of some sort of "sensitivity analysis", meaning simply observation of changes in the results along with the parametrisation over w^K may in this case not only be feasible, but also advised. On the other hand, the weight w^K need not be entirely arbitrary, depending, e.g., on the numbers of variables, their variances or some other pertinent characteristics.

4.8 Minimum Message Length

The minimum message length problem is quite a natural extension to the problems, considered in the immediately preceding sections. Namely, we speak of the model and its precision, trying to establish, for a given set of data, a possibly simple ("short") model that would, at the same time, yield a possibly precise rendition of the data. This problem, initially formulated and analysed by Chris Wallace (Wallace and Boulton, 1968) is represented in popular explanations through the objective function, corresponding to the verbal statement provided here, composed of two parts:

$$\begin{aligned} \text{``message length''} &= -\log_2(P(M \wedge E)) \\ &= -\log_2(P(M)) - \log_2(P(E|M)), \end{aligned} \tag{4.21}$$

where M is the specification of the model (hypothesis), and E—of evidence (data), the formula being derived from the code length expression of the Shannon's information theory in the Bayesian perspective.[7]

It is obvious that in order to proceed in a practical manner, starting from formula (4.21), one needs additional specifications and assumptions, which are usually related to the probabilistic or statistical characterisations of the respective phenomena (like, e.g., the nature of distribution function).

Not only the formula (4.21), along with its interpretation, is quite remindful of the precepts to the bi-partial approach, but, actually, the minimum message length (MML) principles are, from the very beginning, including, indeed, the seminal article of Wallace and Boulton (1968), very frequently applied to the classification and clustering problems. The examples are provided by the studies of Oliver, Baxter and Wallace (1998)—see Sect. 4.3 in this Chapter for the analogous problem, Fitzgibbon, Allison and Dowe (2000), Fitzgibbon, Dowe, and Allison (2002), Davidson (2000), or Asheibi, Stirling, and Soetanto (2008).

Just for the sake of illustration, we shall provide here a very simplistic instance, explaining primarily the meaning of MML and the reasoning behind, this instance being represented by two quite academic cases.

[7]It is common to treat the minimum length problem and approach as (practically) indistinguishable from the minimum description length (MDL) problem formulation, first coined in by Rissanen (1978). The thin distinction—in the opinion of this author—is that MDL looks for the minimum length 'coding' of a definite set of data items, rather than for the simplest and still effective model (and is thus quite analogous to the rule extraction problem of Sect. 4.6). A very telling quotation from Grünwald (1998) on MDL says that it is "based on the following insight: any regularity in a given set of data can be used to compress the data, i.e. to describe it using fewer symbols than needed to describe the data literally". Because of the quite close meaning and possibility of different interpretations, MDL is also used in the clustering context, see, e.g., Figueiredo, Leitão, and Jain (1999), or Böhm et al. (2006), in a way very much like that of MML. Actually, the two examples, provided in this section, can be interpreted in the perspective of both MML and MDL.

Case 1

Assume we deal with our standard data set X and we would like to describe this data set through a minimum length message. On the top of the usual assumptions made here, we include the value ε, being the "threshold of discernibility". This means that any two x and y such that $d(x, y) \leq \varepsilon$ cannot be distinguished (from the point of view of the intended use of the data set). Such an assumption, which is also made in many approaches, is justified by, for instance, the characteristics of the measurement error, or the pragmatic use of the objects, described by the data. The notion of ε is exactly equivalent (in terms of the additional assumptions or information provided) to the previously mentioned necessary additional information on the characteristics of the respective probability distributions or other similar "auxiliary data".

Namely, if $d(x, y) \leq \varepsilon$ then we can replace the descriptions of x and y by a single description, say, that of x, and the indication that we deal with two objects (i.e.: $[x, 2]$), thereby x becoming, in a way, a "representative" of y.

All the descriptions of the separate objects are composed of nm numbers (in the notation, introduced above, actually: $n(m + 1)$ numbers, and we abstract of the potentially various lengths of these numbers here), i.e. n "rows" of the form $[x_i, 1]$, $i = 1, \ldots, n$. Yet, we can group the objects by using the notion of distance and the value of ε. Of course, if

$$\varepsilon < \min_{i,j \in I} d(x_i, x_j)$$

then the whole exercise has no sense.

Thus, we assume that the above is not true, and try to find groups A_q, $q = 1, \ldots, p$, such that possibly many (ideally, of course, all) of the objects in A_q are indistinguishable. We can then shorten the respective descriptions by exploiting the indiscernibility and the possibility of joint description of objects, as roughly illustrated in Fig. 4.16.

In Fig. 4.16 we deal with altogether 15 objects, for which a division into $p = 5$ clusters is suggested. If each object is described separately, the message is composed of $15 \cdot 2 = 30$ numbers ($15 \cdot 3 = 45$ in the notation adopted). The value of ε is illustrated with a circle of corresponding diameter, which allows for the visual assessment of the potential configurations in the particular suggested clusters. It can easily be seen that for some of the exemplary clusters the assignment of objects to representatives is by no means obvious, and constitutes a problem in itself, which shall not be considered here. We assume that we can, and in fact do, obtain the "optimum" such assignment for each of the potential clusters. Suffice to state that it can be seen in Fig. 4.16 that the cluster-wise descriptions might have the following lengths: for $q = 1$: 3; $q = 2$: 3; $q = 3$: 3; $q = 4$: $2 \cdot 3 = 6$; and for $q = 5$: $3 \cdot 3 = 9$, altogether 24 numbers, i.e. less than the original 30.

Case 2

In a certain analogy to Case 1, we might refer to the fuzzy-set-based clustering methods and formulations. It is, namely, quite common for, e.g., the classical FCM algorithm (*fuzzy c-means*, being the fuzzy-set-based counterpart of k-means) that

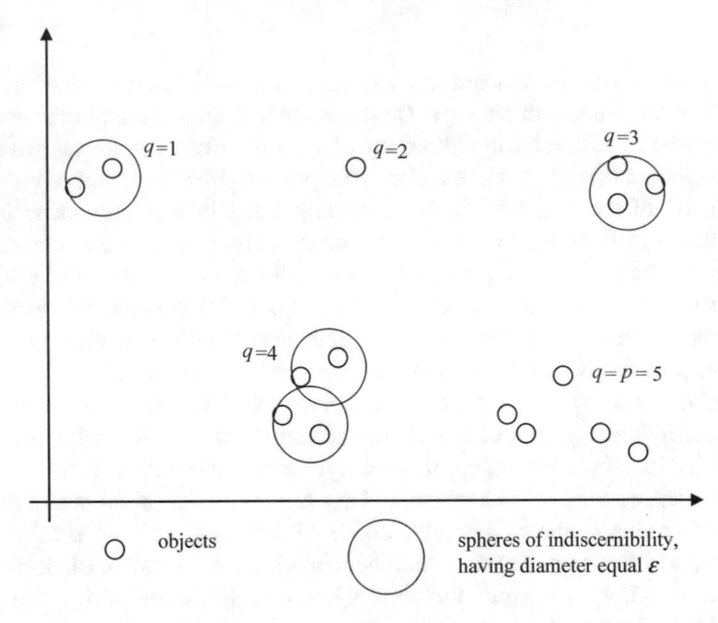

Fig. 4.16 An illustration of various situations arising in connection with indiscernibility applied to MML/MDL type problem of clustering

the membership values of particular objects i in clusters q, $\mu_q(i) \in [0, 1]$, are equal exactly 1 for some objects, 0 for some other ones, and yet between 0 and 1 for the usually small number of the remaining ones (the empty clusters put apart).

Thus, conform to the schematic illustration of Fig. 4.17, for a cluster A_q of cardinality card $A_q = n_q$, we may have n_{q1} objects with $\mu_q(i) = 1$, n_{q0} objects with $\mu_q(i) = 0$, and $n_q - n_{q1} - n_{q0}$ objects with memberships $\mu_q(i) \in (0,1)$. Under certain—practicality—conditions it may be assumed that the n_{q1} objects are "sufficiently adequately" represented by the characteristics of the cluster q, in the perspective of either MML or MDL. It is known that the relation between the numbers of objects n_{q1} and $n_q - n_{q1} - n_{q0}$ for the clusters q depend upon the choice of the exponent, used in FCM algorithm, and it is exactly this exponent that corresponds to the already mentioned assumed "characteristics of the distribution". In any way, we might use FCM, under appropriate assumptions, as a proxy for solving the MML/MDL like problems.

The reference to MML and MDL, which both imply criterion formulations very much in line with the bi-partial approach ("on the one hand…, on the other hand…"), brings us back to the issue of comparison with at least some of the numerous existing clustering quality indices. We shall deal with this issue at some length in the next chapter of the volume, where we present more extensively the various versions of the bi-partial approach in the domain of clustering.

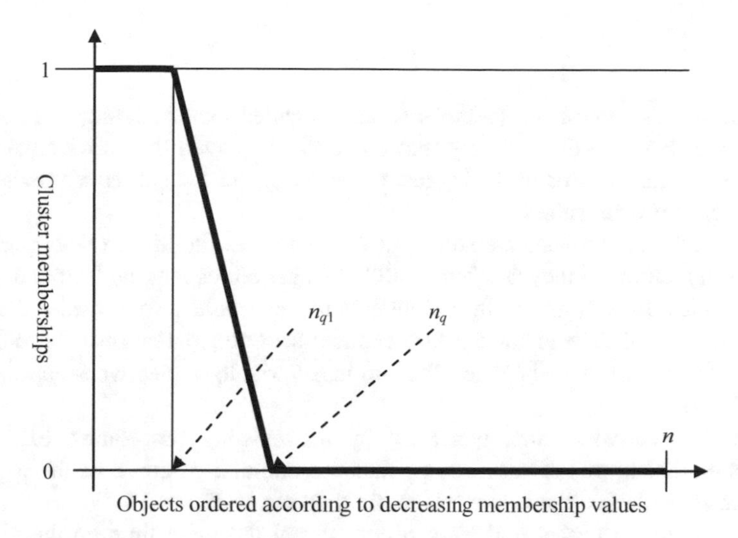

Fig. 4.17 Schematic view of the typical distribution of membership values in a (fuzzy) cluster q, obtained, e.g., with an FCM algorithm

4.9 Number of Factors in Factor Analysis

At this point let us mention—again—that there are, indeed, many problems that can be treated in the very general version of the bi-partial framework. In some cases this is quite trivial, while in other ones it requires quite some effort for the adequate formulation and possibility of obtaining a solution.

An illustration to the point, not yet considered nor mentioned in this chapter, is constituted by the question of the ***number of factors in factor analysis*** (or of principal components in the principal component analysis, these two being very closely related). It is namely, very well known that the actual number of retained factors is chosen on the basis of various criteria, which are either of quite heuristic character, or have little to do with the character and meaning of the concrete data analysed.

In short, in factor analysis (see, e.g., Comrey and Lee, 1992, or Hair, Tatham, Anderson, & Black, 1998) we wish to obtain, from our data set X, here perceived along columns-variables, $x_{.k}$, $x_{.k} = \{x_{ik}\}_I$, a set of variables y_l, $l = 1, \ldots, p$, such that: the variables y_l are pairwise orthogonal (uncorrelated), and we try to obtain a possibly small number p of these variables that would "explain the possible biggest share of the variance in X". These new variables are called factors. Of course, when $p = m$, then we can obtain complete explanation of the said variance.

The factors y_l are identified as linear combinations of the variables $x_{.k}$:

$$y_l = a_l^T X \tag{4.22}$$

where a_l are the vectors of coefficients, the so-called factor loadings. In obtaining the a_l, and therefore the y_l a very important role is played by matrix eigenvalues. Actually, y_1 corresponds to the biggest eigenvalue, and the subsequent factors—to the decreasing eigenvalues.

In fact, the eigenvalues are sometimes indicated as the hint to determining the number of factors—if they are "too small", then the corresponding factors ought not be accounted for any more. In a similar vein one could also consider the factor loadings, since they tend for the further factors to be smaller and dispersed (this being indicative especially when the original variable values were appropriately normalised).

The consecutive factors, generated by the existing procedures, explain the decreasing shares of the variance, so that the cumulative curve of the explained variance share looks like in the schematic diagram of Fig. 4.18.

By exploiting the fact that we can consider at the same time, in the bi-partial framework, the proportion of variance explained and the "clarity" or "strength" of the factors, it becomes possible to avoid using the one-sided rules-of-the thumb, like the "elbow" on the curve of explained variance, or the eigenvalue exceeding one (see, for instance, Hair et al., 1998, or Guadagnoli & Velicer, 1988). The concrete measures used and their parameters allow also for accounting for the scale involved, often associated with the nature of the data (e.g. our knowledge of their precision and degree of representativeness for the problem studied).

In the subsequent section, on the other hand, we give some hints on another problem, which appears to have even less in common with the principles of the bi-partial approach, and which will be treated also more extensively later on in the book.

4.10 Ordering—Preference Aggregation

Assume, as before, that we deal with objects indexed i, $i = 1, \ldots, n$, for which we dispose of the multidimensional data, describing these objects, $x_i = \{x_{i1}, x_{i2}, \ldots, x_{ik}, \ldots, x_{im}\}$. We wish to order these objects, and, without any loss to generality, we adopt the rule that bigger values precede smaller values, so that $x_{ik} > x_{jk} \Leftrightarrow x_{ik} \diamond x_{jk}$, where $\diamond$ denotes the precedence of the left argument before the right argument. (We also assume that the same implication will apply, when $>$ is replaced by $\geq$.)

When $x_{ik} \geq x_{jk}$ for all $k = 1, \ldots, m$, then we can deduce that $x_i \diamond x_j$. However, when some x_{ik} are bigger than x_{jk}, and some are smaller, we cannot proceed with such a simple reasoning. This issue, which constitutes in itself a very broad domain, not just of data analysis, but also "social choice" and related domains, is being resolved in a very wide variety of manners. One of the most popular, and, indeed,

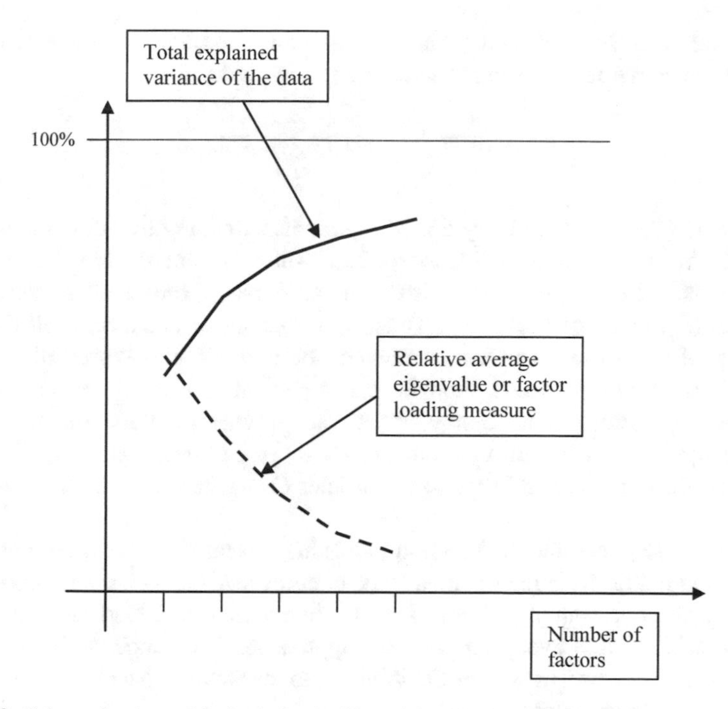

Fig. 4.18 Schematic view of the basis for the bi-partial setting to solve the problem of the number of factors in factor analysis

quite intuitive and effective, ways to deal with this issue is to assign weights w_k to particular descriptive variables, with $w_k \in [0, 1]$, and frequently also $\Sigma_k w_k = 1$, and then to compare the numbers $\Sigma_k w_k x_{ik}$ rather than the vectors x_i (and it is, of course, also always possible to take all weights equal $1/m$).

This issue becomes even more intricate when some comparisons x_{ik} vs. x_{jk} cannot be made, or expressed, for some reason (lack of measurements, first of all), and such a situation is not systematically repeated for all objects. Yet another situation arises when the comparisons x_{ik} vs. x_{jk} take the form of numbers that exceed the three-valued representation of the cases (1) $x_{ik} > x_{jk}$; (2) $x_{ik} = x_{jk}$; (3) $x_{ik} < x_{jk}$. This may be, for instance, the situation of m experts, providing assessments of preference for the pairs (i, j), these assessments taking the form of, say, scores between −1 and +1.[8]

The approach that we propose for tackling this type of issues, in line with the bi-partial paradigm, consists in the formulation of the global objective function, accounting for all the objects and the entire ordering, with simultaneous

[8]Another case of interest, which can be effectively treated in the framework here proposed, is the one of a sports tournament, in which teams play in pairs, achieving various scores in their matches, but not all pairs of teams actually meet and play with each other.

consideration of the "for" and "against", regarding the particular pairs of objects. Such an objective function might have the following form:

$$Q_{PA} = \sum_{i,j \in I} y_{ij}\delta_{ij} + \sum_{i,j \in I} y_{ji}\sigma_{ij} \quad (4.22)$$

where $y_{ij} \in \{0,1\}$ are the binary decision variables, defining the ultimate ordering, resulting from the maximisation of (4.22), while δ_{ij} and σ_{ij} are the summary expressions, reflecting the total preference of i over j across all m description variables (δ_{ij}) and vice versa (σ_{ij}). Thus, $\delta_{ij} + \sigma_{ij}$ may be equal m, if all the comparisons of x_{ik} vs. x_{jk} can be accounted for and all are expressed as sharp inequalities. We may assume, though, that in general $\delta_{ij} + \sigma_{ij} \leq m$. The decision variables y_{ij}, besides being binary, must also provide for the ultimate ordering (transitivity). We shall return to this subject in greater detail in Chap. 7, and an interested Reader is also referred to the articles Owsiński (2011, 2012a), with this respect.

Suffice to say here that the function (4.22) has a clear bi-partial tang. Let us add that although this formulation in a way circumvents the weighing problem by implicitly assuming equal weights of descriptive variables, it also offers a number of distinct advantages over many existing approaches. They include the robustness against missing values, as well as flexibility as to the way the "scores" δ_{ij} and σ_{ij} are calculated and represented.

A different issue is the algorithmic one, and we shall return to it at greater length in Chap. 7.

References

Agrawal, R., Imielinski, T., & Swami, A. (1993). Mining association rules between sets of items in large databases. In *SIGMOD*, vol. 5. Washington, DC: ACM Press, pp. 207–216.

Asheibi, A., Stirling, D., & Soetanto, D. (2008). Determination of the optimal number of clusters in harmonic data classification. In *ICHQP 2008: 13th International Conference on Harmonics & Quality of Power*. IEEE Publications.

Bock, H.-H. (1994). Classification and clustering: Problems for the future. In E. Diday, et al. (Eds.), *New approaches in classification and data analysis* (pp. 3–24). Berlin: Springer.

Böhm, C., Faloutsos, Ch., Pan, J.-Y., & Plant, C. (2006). Robust information-theoretic clustering. In *KDD'06*, August 20–23, 2006. Philadelphia, Pennsylvania, USA: ACM Press.

Bramer, M. (2007). *Principles of data mining*. New York: Springer.

Comrey, A. L., & Lee, H. B. (1992). *A first course in factor analysis* (2nd ed.). Hillsdale, N.J.: Lawrence Erlbaum Associates.

Czekanowski, J. (1909). Zur Differenzialdiagnose der Neandertal-gruppe. *Korrespondenz-Blatt der Deutschen Gesellschaft für Anthropologie etc. 1*, 40.

Czekanowski, J. (1913). Zarys metod statystycznych w zastosowaniu do antropologii (An outline for the statistical methods in application to anthropology; in Polish, Vol. 5). *Warszawa: Prace Tow. Nauk. Warsz., III Wydz. Nauk Mat. i Przyr.*

Czekanowski, J. (1926). Metoda podobieństwa w zastosowaniu do badań psychometrycznych (The method of similarity applied to psychometric studies; in Polish, Vol. 3). *PTF, Badania Psychologiczne*, Lwów.

Czekanowski, J. (1932). Coefficient of racial likeness and "durchschnittliche Differenz". *Anthrop. Anz.* 9.

Davidson, I. (2000). Minimum message length clustering using Gibbs sampling. In *The 16th International Conference on Uncertainty in Artificial Intelligence.* Stanford University.

Figueiredo, M. A. T., Leitão, J. M. N., & Jain A. K. (1999). On fitting mixture models. In E. R. Hancock & M. Pelillo (Eds.), *Energy minimization methods in computer vision and pattern recognition. EMMCVPR 1999. Lecture Notes in Computer Science*, 1654. Berlin, Heidelberg: Springer.

Fitzgibbon, L. J., Allison, L., & Dowe, D. L. (2000). Minimum message length grouping of ordered data. In: H. Arimura, S. Jain, & A. Dharma (Eds.), *Algorithmic learning theory. ALT 2000. Lecture Notes in Computer Science*, 1968. Springer, Berlin, Heidelberg.

Fitzgibbon, L. J., Dowe, D. L., & Allison, L. (2002). Change-point estimation using new minimum message length approximation. In M. Ishizuka & A. Sattar (Eds.), *PRICAI 2002*. LNAI 2417. Berlin-Heidelberg: Springer, pp. 244–254.

Gan, G., Ma, C., & Wu, J. (2007). *Data clustering. Theory, algorithms and applications.* Philadelphia: SIAM & ASA.

Greco, S., Słowiński, R., & Szczęch, I. (2009). Analysis of monotonicity properties of some rule interestingness measures. *Control & Cybernetics, 38*(1), 9–25.

Grünwald, P. (1998). *MDL Tutorial.* Retrieved December 12, 2017 from http://homepages.cwi.nl/~pdg/ftp/mdlintro.

Guadagnoli, E., & Velicer, W. (1988). Relation of sample size to the stability of component patterns. *Psychological Bulletin, 103,* 265–275.

Hair, J. F., Tatham, R. L., Anderson, R. E., & Black, W. (1998). *Multivariate data analysis* (5th ed.). London: Prentice Hall.

Hansen, P., Brimberg, J., Urosević, D., & Mladenović, N. (2009). Solving large p-median clustering problems by primal-dual variable neighbourhood search. *Data Mining and Knowledge Discovery, 19,* 351–375.

Hilderman, R., & Hamilton, H. (2001). *Knowledge discovery and measures of interest.* Boston: Kluwer.

Liao, K., & Guo, D. (2008). A clustering-based approach to the capacitated facility location problem. *Transactions in GIS, 12*(3), 323–339.

Mulvey, J. M., & Beck, M. P. (1984). Solving capacitated clustering problems. *European Journal of Operational Research, 18,* 339–348.

Nielsen, L. (2011). *Classification of Countries based on their level of development: How it is done and how it could be done.* IMF Working Paper, WP/11/31, IMF.

Oliver, J. J., Baxter, R. A., & Wallace, C. S. (1998). Minimum message length segmentation. In X Wu, R. Kotagiri, & K. B. Korb (Eds.), *Research and development in knowledge discovery and data mining. PAKDD 1998. Lecture Notes in Computer Science (Lecture Notes in Artificial Intelligence)*, 1394. Berlin, Heidelberg: Springer.

Owsiński, J. W. (2001). Clustering as a model and an approach in flexible manufacturing. *Taksonomia 8. Klasyfikacja i analiza danych – teoria i zastosowania*, In K. Jajuga & M. Walesiak, (Eds.), *Prace Naukowe AE we Wrocławiu* (No. 906, pp. 168–179). Wrocław: Wydawnictwo AE we Wrocławiu.

Owsiński, J. W. (2009). Machine-part grouping and cluster analysis: Similarities, distances and grouping criteria. *Bulletin of the Polish Academy of Sciences. Technical Sciences.* Special Issue: *Modeling and Optimization of Manufacturing Systems*, *57*(3), 217–228 (Guest Editors: Z. A. Banaszak, J. Józefczyk).

Owsiński, J. W. (2011). The bi-partial approach in clustering and ordering: the model and the algorithms. *Statistica & Applicazioni*, 43–59 (Special Issue).

Owsiński. J. W. (2012a). Clustering and ordering via the bi-partial approach: the rationale, the model and some algorithmic considerations. In J. Pociecha & Reinhold Decker (Eds.), *Data analysis methods and its applications* (pp. 109–124). Warszawa: Wydawnictwo C.H. Beck.

Owsiński, J. W. (2012b, June). *On dividing an empirical distribution into optimal segments*. Rome: SIS (Italian Statistical Society) Scientific Meeting. http://meetings.sis-statistica.org/index.php/sm/sm2012/paper/viewFile/2368/229.

Owsiński, J. W. (2012c). On the optimal division of an empirical distribution (and some related problems). *Przegląd Statystyczny, 1,* 109–122. (Special issue).

Owsiński, J. W., Stańczak, J., Sęp, K., & Potrzebowski, H. (2010). Machine-part grouping in flexible manufacturing: Formalisation and the use of genetic algorithms. In P. Leitão, C. E. Pereira, J. Barata, (Eds.), *10th IFAC Workshop on Intelligent Manufacturing Systems* (pp. 233–238). IFAC (DVD).

Owsiński, J. W., & Tarchalski, T. (2008). Pomiar jakości życia. Uwagi na marginesie pewnego rankingu (Measuring life quality. Remarks relative to a certain ranking; in Polish, No. 1, pp. 59–95). *Zeszyty Naukowe Wydziału Informatycznych Technik Zarządzania "Współczesne Problemy Zarządzania"*.

Piatetsky-Shapiro, G. (1991). Discovery, analysis and presentation of strong rules. *Knowledge Discovery in Databases, 2,* 29–248.

Rissanen, J. (1978). Modeling by shortest data description. *Automatica, 14*(5), 465–471.

Wallace, C. S., & Boulton, D. M. (1968). An information measure for classification. *Computer Journal, 11*(2), 185–194.

Chapter 5
Formulations in Cluster Analysis

In this chapter we shall present and analyse in a bit more of detail the examples of application of the bi-partial approach in the domain, from which it actually arose, that is—from cluster analysis. We shall start from the "leading example" that was presented in Chap. 3, Sect. 3.2. Then, we shall present, in a relatively extensive treatment, the bi-partial version of the well known k-means algorithm, and a couple of other potentially applicable versions of the bi-partial clustering formulations.

5.1 The Leading Example Again

5.1.1 The Formulation Repeated

We quote here explicitly, with the comments, resulting from the examples considered so far, the series of formulae from Chap. 3 (preserving even their numbers), which represent, as said there, possibly the simplest and the most intuitive version of the bi-partial clustering formulations. It has the following general form[1]:

$$\begin{aligned} Q^D{}_S(P) &= C_S(P) + C^D(P) \\ &= \sum_q \sum_{i<j \in A_q} s_{ij} + \sum_q \sum_{q' > q} \sum_{i \in A_q} \sum_{j \in A_{q'}} d_{ij}. \end{aligned} \tag{3.3}$$

In this manner we account in the maximised objective function (3.3) for all the proximities between the objects, forming the particular clusters, and for all the distances between the objects that are placed in different clusters. There are no other notions involved, and the sole preoccupation appears to be associated with the

[1]We retain here the formulae numbering from the original parts of the text, where they appear for the first time.

J. W. Owsiński, *Data Analysis in Bi-partial Perspective: Clustering and Beyond*, Studies in Computational Intelligence 818,
https://doi.org/10.1007/978-3-030-13389-4_5

relation between d_{ij} and s_{ij}. It should be noted, at this point, that d_{ij} and s_{ij} in this formulation are defined for two disjoint sets of pairs of objects, which allows for a relatively wide margin of definitions regarding the relation between these two notions, in particular—separate definitions of the two.

Quite a "natural" definition of the transformation between d and s is

$$s(d) = d^{\max} - d, \tag{3.4}$$

where $d^{\max} = \max_{i,j} d_{ij}$ is the diameter of the set of objects, X, which, after normalisation, when $d_{ij} \in [0,1]$ $\forall$ $i, j \in I$, takes the form of

$$s(d) = 1 - d. \tag{3.5}$$

Like in the case of other forms of the bi-partial objective function, we are especially interested in the properties of the function $Q_D{}^S(P)$ and its components, $C_S(P)$ and $C^D(P)$, for the sequences of nested partitions, $\{P^t\}_t$, that is, such that at least one of the clusters $A_q{}^t$ in partition P^t is the sum of at least two clusters $A_q{}^{t-1}$ from the preceding partition P^{t-1}, all the other clusters remaining the same for the two partitions. This sequence starts with P^0, of which we assume that the clusters $A_q{}^0$, forming it, are equivalent to individual objects, so that $p = n$. The sequence ends with some P^T, where $T \leq n - 1$, this terminal partition being composed of just one all-embracing cluster, i.e. $p = 1$, $A_1 = I$.

Thus, for the formulation (3.3) it is easily seen that in the case of partition P^T

$$Q^D{}_S(P^T) = C_S(P^T) = \sum_{i<j \in I} s_{ij} = S(I), \tag{3.6}$$

which, for the definition of $s(.)$ as in (3.4), turns into

$$Q^D{}_S(P^T) = S(I) = d^{\max} n(n-1)/2 - D(I).$$

Now, if we deal with partition P^0, which is composed of all the objects being separate clusters, then

$$Q^D{}_S(P^0) = C^D(P^0) = \sum_{i<j \in I'} d_{ij} = D(I). \tag{3.7}$$

When comparing (3.6) with (3.7) we may come to the conclusion that it is reasonable to apply yet another definition of $s(.)$, meant to secure that $Q_D{}^S(P^T) = Q_D{}^S(P^0)$), namely the equality of the objective function values at the two extremes of the possible partitions. The function $s(.)$, which satisfies this condition, and at the same time another reasonable condition of $s^{\min} = d^{\min}$ (generally stronger than the necessary $s^{\min} \geq 0$), has the form

$$s(d) = d^{average} \frac{d^{\max} - d^{\min}}{d^{\max} - d^{average}} - \frac{d^{average} - d^{\min}}{d^{\max} - d^{average}} d. \tag{3.8}$$

In Chap. 3 some attention was devoted to the transformation $s(d)$, in view of its key role in the bi-partial approach for clustering. Thus, in particular, one could also reason for the use of the function $s(.)$ that preserves the range of values of d, i.e.

$$s^{\max} - s^{\min} = d^{\max} - d^{\min},$$

requiring, again, additionally, $s^{\min} = d^{\min}$, which is equivalent to a simple definition

$$s(d) = d^{\min} + d^{\max} - d. \tag{3.9}$$

Yet, notwithstanding the form of $s(d)$ that we choose, it is easy to show that for the definition (3.3) the two components of $Q_D{}^S(P)$, namely $C_S(P)$ and $C^D(P)$, display opposite monotonicity along the sequences of nested partitions, $\{P^t\}_t$. We shall yet be showing some further reaching properties of this formulation.

5.1.2 The Reference to the Relational Form of the Bi-partial Objective Function and the Respective Solution

The bi-partial form of the objective function, recalled here, namely

$$Q^D{}_S(P) = \sum_q \sum_{i<j\in Aq} s_{ij} + \sum_q \sum_{q'>q} \sum_{i\in Aq} \sum_{j\in Aq'} d_{ij}.$$

had been proposed already before by Marcotorchino and Michaud (1979, 1982), see also Owsiński (2011, 2012), in their broader relational scheme of data analysis. The respective original proposal from Marcotorchino and Michaud called for obtaining of the clustering solution with the use of the mathematical programming systems and softwares, the respective mathematical programming (MP) problem formulation being as follows:

$$\text{maximise} \sum_{i,j\in I} \left(y_{ij}s_{ij} + (1 - y_{ij})d_{ij}\right) \tag{5.1}$$

where $y_{ij} = 1$ when objects i and j belong to the same cluster, and $y_{ij} = 0$ when they belong to different clusters, this formulation being subject to the following additional constraints:

$$
\begin{aligned}
&y_{ij} \in \{0,1\}, \quad \forall i,j,\\
&y_{ij} = y_{ji}, \quad \forall i,j, \text{meaning symmetry}\\
&y_{ij} + y_{jv} - y_{iv} \le 1, \quad \forall i,j,v, \text{meaning transitivity.}
\end{aligned} \tag{5.2}
$$

This formulation leads to a relatively simple parameterised LP problem, in which the "proximity component" is weighted against the "distance component" in the objective function (5.1), given above. Thus, for the objective function (5.1), the constraints (5.2), and the normalised distance/proximity values ($d_{ij} = 1 - s_{ij}$) we can formulate the LP problem, essentially equivalent to the one below:

$$
\max_P Q(P,r) = r \sum_q \sum_{i<j \in Aq} s_{ij} + (1-r) \sum_{q<q'} \sum_{i \in Aq} \sum_{j \in Aq'} d_{ij} \tag{3.10}
$$

where the parameter r, reflecting the weights of the components, associated with the intra-cluster similarity and the inter-cluster dissimilarity (quite intuitively supposed to take the value of 1/2 in the proper solution to the original problem), ranges from 1 (full weight on intra-cluster coherence) to 0 (full weight on inter-cluster dissimilarities, that is—distances).

By solving the problem (3.10), (5.2) for a sequence of values of r from 0 upwards to 1, one obtains a sequence of partitions, ranging between two extremes, determined by the disappearance of one of the right hand side components in (3.10). A bit more light on this sequence, already virtually without reference to the LP formulation, shall be forwarded in the next section.

5.1.3 The Prerequisites for the Algorithm Relative to the MP Formulation

Thus, as said, once we adopt the objective function in the form of (3.10) with the accompanying constraints as in (5.2), we could run, for definite values of $r \in [0, 1]$, the LP programs that would yield the optimum values of y_{ij}, meaning the optimum partition P of the given set of objects.

It is extremely interesting and informative to watch the solutions of the problem (3.10) changing with the value of the parameter r.

Thus, for $r = 0$, the formulation of the problem implied by (3.10), given no other constraints, especially regarding the number of clusters, except for those given in (5.2), enforcing the partition, would yield the solution, in which each of the objects would constitute a separate cluster, i.e. $p = n$, since the first component in (3.10) simply disappears. Given the constraints (5.2), such a solution is feasible, and shall indeed be provided by any (correct) method whatsoever.

As the value of the parameter r is increased from 0, the very first encountered obvious solution to (3.10), satisfying all of the constraints (5.2), and different from the one, obtained for $r = 0$, would appear to be the one, which merges two most similar (least distant) objects i^* and j^* (i.e. such that $d_{i*j*} = \min_{i,j} d_{ij}$). The switch from the optimum partition for $r = 0$, which we shall denote $P^*(r = 0) = P^*(0) = P^0$, which consists in all objects being separate clusters, $P^0 = I$, to the one, in which objects indexed i^* and j^* form a two-object cluster, takes place at a definite value of the parameter $r > 0$. Denote this value, at which P^0 is replaced by P^1, defined by the merger of i^* and j^*, by, accordingly, r^1 (implying that $r^0 = 0$).

Let us note, as a side remark, that, quite obviously, if there exist in the data set the pairs (i, j), for which $d_{ij} = 0$, then the value of (3.10) and (5.1) does not change, whether we merge these objects, or not. Actually, we can assume that for such cases $r^1 = r^0 = 0$, with r^1 corresponding to P^1, the partition, which incorporates the merger of all the objects, among which all the distances are zero. An analogous reasoning applies when there are more equidistant objects than just pairs at the level of minimum distance.

If, however, there are no (more) pairs (i, j), for which $d_{ij} = 0$, then the merger of a pair i^* and j^* takes place for the parameter value r^1 determined by d_{i*j*} and s_{i*j*}. In order to show the relation, take the two partitions, P^0 and P^1, the latter one differing from the former by just one merger, and the values of objective function for these partitions, $Q(P^0, r)$ and $Q(P^1, r)$. Hence, we compare the values of

$$Q(P^0, r) = r\sum_i s_{ii} + (1-r)\sum_{i<j} d_{ij} \tag{5.3}$$

and

$$Q(P^1, r) = r(\sum_i s_{ii} + s_{i^*j^*}) + (1-r)(\sum_{i<j} d_{ij} - d_{i^*j^*}) \tag{5.4}$$

for the parameter r increasing up from 0. Obviously, the values of $\sum_i s_{ii}$ $(=S(I))$ and $\sum_{i<j} d_{ij}$ $(=D(I))$ are constants for the given set of objects. The comparison of $Q(P^0, r)$ and $Q(P^1, r)$ yields:

$$r\sum_i s_{ii} + (1-r)\sum_{i<j} d_{ij} = r\left(\sum_i s_{ii} + s_{i^*j^*}\right) + (1-r)\left(\sum_{i<j} d_{ij} - d_{i^*j^*}\right) \tag{5.5}$$

hence, after simple operations, yielding

$$rs_{i^*j^*} - (1-r)d_{i^*j^*} = 0 \tag{5.6}$$

we get the sought value of r^1:

$$r^1 = \frac{d_{i^*j^*}}{d_{i^*j^*} + s_{i^*j^*}}. \tag{5.7}$$

As we remember that the pair (i^*, j^*) corresponded to the smallest distance (and so, at the same time, the biggest proximity, i.e. similarity), the expression (5.7) definitely leads to the smallest value of r among those that might arise from the merger of any pair of objects in the set considered.

Since we shall yet return in an ampler manner to the issue of the algorithm in Chap. 6, only a couple of remarks shall be forwarded here.

First, it is quite straightforward to observe that the reasoning, leading to formula (5.7), and the simile of this formula, obtained for the merger operation under the objective function considered, as well, apply to (i) all the subsequent (disjoint) pairs of objects in the set X, ranked according to their pairwise distances, and (ii) all the subsequent mergers of clusters, no matter how many objects they may contain (in this case the respective formula would account for the appropriate characteristics of clusters and pairs of clusters involved).

Second, it should be noted that formula (5.7) provides a simple rule for merging the objects (and clusters), very much reminding the rules that are used by the classical agglomerative schemes (in fact, it is equivalent to one of such rules). Yet, in this case we do not only proceed as in these schemes, guided by a definite merger rule, but dispose also of the "global" objective function, which allows for the evaluation of the consecutive partitions obtained. The approach offers two options with this respect, namely:

1. to track the values of $Q(P, r^t)$ along the consecutive mergers and then to pick the partition, corresponding to the maximum value of the objective function (this option calls for a definite post factum interpretation of the value of r, corresponding to the solution, which might not be quite straightforward);
2. to simply stop after the value of $r = 1/2$ has been exceeded, this second option being especially justified by an appropriate scaling of $s(d)$.

Third, the algorithmic scheme here outlined leads, definitely, to suboptimisation, since only mergers are performed, and there is no evidence that this secures reaching the optimum solution. In Chap. 6 this issue shall be considered in somewhat greater detail. Now, it suffices to mention that if the LP problem (5.1)–(5.2) were solved directly, for the varying values of the parameter r, then it could be verified whether, to what extent, and how, for the particular data sets, the proposed agglomerative scheme fails to find the optimum partitions.[2]

[2]For a limited number of academic examples such a comparison was done by this author, together with Sławomir Zadrożny, and for this sample the results were the same.

5.2 The Bi-partial Version of the k-means Algorithm

5.2.1 The Standard Case

This section is devoted to presentation of the bi-partial version of the classical k-means algorithm. The version here introduced and described differs from the original one by the formulation of the objective function, with the consequences, concerning the nature of solutions obtained, and, most importantly, the possibility of specifying the number of clusters, this being routinely done for the classical k-means only through the use of "external" criteria.

Namely, let us remind that the general formulation of the classical k-means is based on the minimised objective function that we shall denote in the usual manner here adopted, $C_D(P)$. In the standard setting of the classical k-means algorithm, $C_D(P)$ has the form

$$C_D(P) = \sum_q \sum_{i \in Aq} d(x_i, x^q), \tag{5.8}$$

where $d(.,.)$ is some distance function,[3] x_i is a vector of values, characterising object i, and x^q is the (analogous) characterisation of the "representative" of cluster A_q, $q = 1, \ldots, p$.

The minimum values of $C_D(P)$ are obtained with the use of the primeval generic algorithm associated with the k-means, namely the "centre-and-reallocate" one:

> for some initial, possibly random, set of x^q, $q = 1, \ldots, p$, assign the elements of the set X to the closest x^q, determining thereby the clusters A_q, then calculate the new x^q, e.g. as averages over A_q, and go back to the assignment step, stopping the entire procedure whenever there is no change between the iterations, or the change satisfies definite conditions.

It is known that this algorithm very efficiently leads to a local minimum of $C_D(P)$—in a very limited number of iterations for even quite large data sets.

In addition, even though it is known that the "centre-and-reallocate" procedure attains just a local minimum of the given objective function, starting the procedure from multiple randomly generated sets of "representatives" x^q ultimately yields the proper minimum for the given number of clusters p.

The minimum values of $C_D(P)$, determined with the primeval algorithm for consecutive numbers of clusters, p, these values denoted here as $C^*{}_D(p)$, decrease, from a simile of total variance, conform to (5.8), for $p = 1$, i.e. for the entire set of objects considered, treated as a single cluster, down to zero for $p = n$ (or even "earlier", in terms of p). Hence, it is necessary to run the respective algorithm, with

---

[3]Actually, in order to ensure the fulfilment of the appropriate properties of the k-means algorithm, especially related to the securing that the cluster mean minimises the sum of distances to the objects in a cluster, mostly squared Euclidean distance is referred to in this context. Otherwise, the basic properties mentioned are replaced by respective approximations.

repetitions, mentioned before, for several consecutive values of p and apply an external (statistical) criterion, say, AIC, BIC, Calinski & Harabasz, etc., in order to pick the "proper" number of clusters p.

In this manner we are obliged, in order to find the "best" p, to recur to a criterion that does not stem from the same "philosophy" as the objective function $C_D(P)$ and the respective algorithm, and thus a criterion applied may be poorly fit to the original k-means framework. This important reservation holds true even if we accept the results produced by such an "external" criterion on the basis of comparison of results produced by several different criteria (all of them being actually in principle similarly "external" to the original procedure).

The rationale behind the modification along the lines of bi-partial objective function is to have a clear intuitive tool for selecting the "proper" p, while preserving the simplicity of the original approach, and without the necessity of referring to "external" criteria, which may not have a direct connection to the very algorithm and its basic precepts. Yet, the issue is, actually, more general, since the possibility of obtaining the "proper" p is just a by-product of the fact that the form of the bi-partial objective function indeed models the basic formulation of the clustering problem ("to cluster together the similar, or the close objects, while separating those farther away from each other").

The weak point of the general bi-partial approach is that—quite in line with the remark above—there is no ready recipe for designing the concrete forms of $Q_D{}^S(P)$. It is quite often so that—like in the case of k-means type of algorithms—some objective function, whether explicit or implicit, corresponding to an existing approach, and representing either $C^S(P)$ or $C_D(P)$ ($C^D(P)$ or $C_S(P)$ in the case of the dual formulation), can be complemented with an appropriate counterpart, which has then to be cleverly designed.

The reward, however, may be worth the effort: for the pair of functions fulfilling certain additional conditions, which in some cases are quite natural, one obtains also a straightforward, even if not always very effective, algorithm, leading to the optimum or sub-optimum P. This is, in particular, and that in quite a straightforward manner, the case of the k-means-related version.

In this particular case, the respective objective function, matching the principles of k-means, might have the following minimised form:

$$Q_D{}^S(P) = C_D(P) + C^S(P), \tag{5.9}$$

with $C_D(P)$ defined, for instance, as in formula (5.8). The second component, $C^S(P)$, would then have to reflect the overall measure of inter-cluster proximity (similarity). This measure might take, in particular, the form of

$$C^S(P) = 1/2 \sum_q S^*(A_q), \tag{5.10}$$

with $S^*(A_q)$ being the "outer similarity" measure for the cluster A_q, defined, say, as

$$S^*\left(A_q\right) = \sum_{i \in Aq} \max_{q' \neq q} s\left(x_i, x^{q'}\right). \tag{5.11}$$

5.2.2 *Two Illustrative Examples*

With the definitions provided through the formulae (5.9)–(5.11) a micro-study, carried out for illustrative purposes, was complemented by choosing the definition of distance (here: Manhattan, or city-block, meaning that we deal in the "centering" step only with an approximation of the minimum sum of distances) and the transformation $s(d)$, here—according to the formula (3.8), repeated in this chapter in Sect. 5.1.1 ("average-preserving"). A series of computations were carried out for several simple academic data sets, meant to simply verify the basic features of the proposed objective function.

Two of such data sets, which were used in the study, are shown in Figs. 5.1 and 5.2. It can be easily seen that they are meant to check the possibility of indicating more than one "best" partition, although in the case of Fig. 5.1 the respective (alternative/potential) partitions are well visible, while in Fig. 5.2 the case is not so

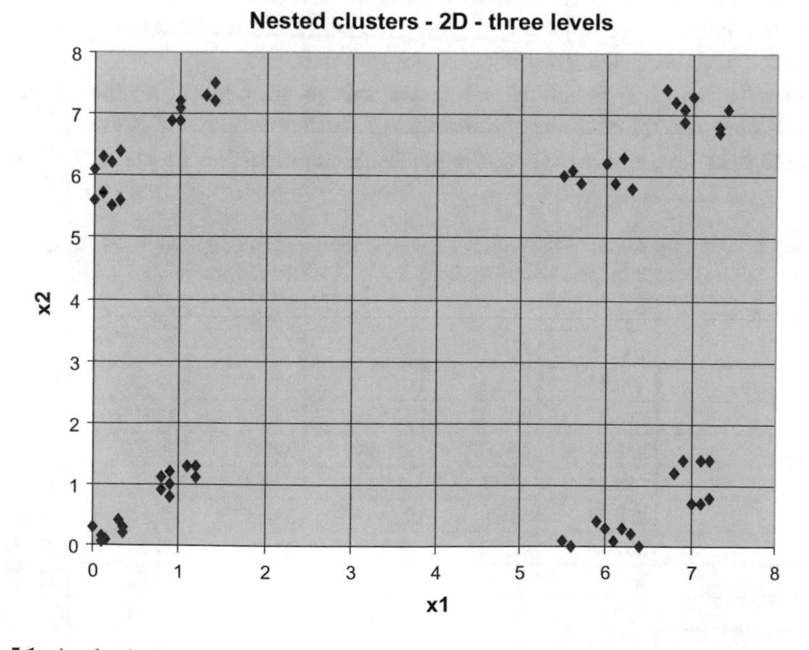

Fig. 5.1 Academic Example 1 of a data set for the bi-partial k-means

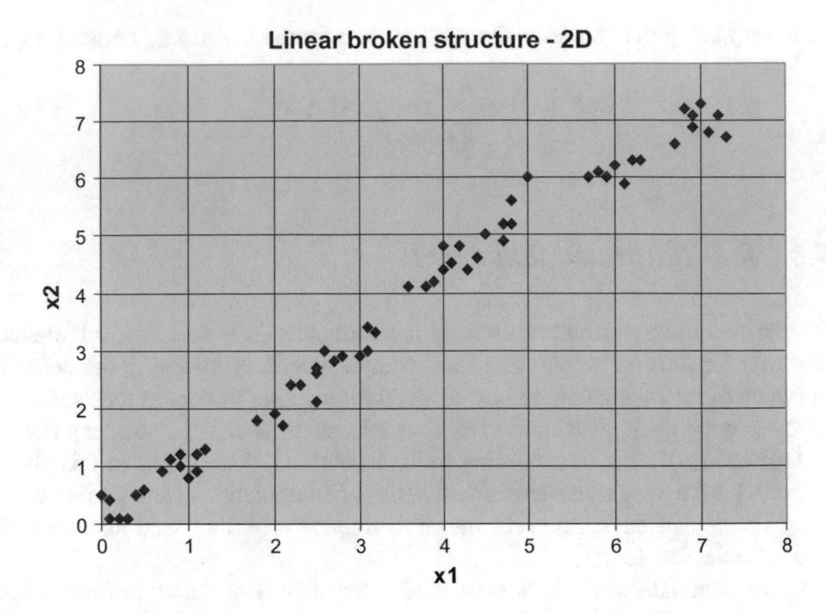

Fig. 5.2 Academic Example 2 of a data set for the bi-partial k-means

evident. It can be said that in the case of Fig. 5.1 one deals with true "nested" partitions (two or even three levels of quasi-optimum partitions).

The results, meaning the values of the functions, involved in (5.9), obtained in respective consecutive solutions for the successive values of p, are provided in Table 5.1, following the illustration of the first data set.

Some technical explanations are due at this point, concerning the exemplary calculations, in addition to the specification of the definitions used in them. Namely, for both data sets, n = 60. Then, the values of the objective function $Q_D{}^S$ for the

Table 5.1 Optimum values of the objective function (5.9) and its components for the successive numbers of clusters for the academic Examples 1 and 2, illustrated in Figs. 5.1 and 5.2

No. of clusters	Example 1 (Fig. 5.1)			Example 2 (Fig. 5.2)		
	$Q_D{}^S$	C_D	C^S	$Q_D{}^S$	C_D	C^S
1	353.65	353.65	0	235.70	235.70	0
2	338.87	207.05	131.82	228.24	117.75	110.49
3	333.67	134.87	198.80	242.25	76.62	165.63
4	265.15	59.67	205.48	234.91	47.99	186.92
5	288.23	51.05	237.18	242.37	42.80	199.58
6	307.38	42.52	264.86	243.72	32.62	211.10
7	X	X	X	245.52	27.84	217.68
8	335.40	21.96	324.44	245.39	22.98	222.41
60	361.23	0	361.23	245.31	0	245.31

extreme numbers of clusters (1 and 60) differ somewhat, despite the application of the "averaging" definition of *s(d)*, due to the roundings, appearing in the actual practical definition of this transformation. Finally, for Example 1 the values for $p = 7$ were not calculated.

Concerning the results obtained in this micro study, it should be admitted that the bi-partial objective function, as designed for this particular case, indicates "correct" values of p as corresponding to the true partitions (clusterings), visible in the data sets. This is particularly true for the Example 1, where a very clear minimum is observed for $p = 4$. No proper local minima, though, are observed, as it could be hoped, for $p = 8$, and possibly higher values of p. Let us add, however, that for $p = 8$ a slight deviation is observed in the respective curve of $Q_D{}^S(p)$ from the upward trend towards $Q_D{}^S(60)$.

Now, regarding Example 2, it can be easily seen, as this was already indicated, that this case is not so evident, also to the human eye and brain, as that of Example 1 and the relative flatness of the line, defined by the values of $Q_D{}^S(p)$ is justified. Yet, also in this case, the distinct local minima were obtained for the values of p equal 2 and then 4, as this could be expected.

So, altogether, the verification of applicability of the bi-partial approach to the k-means-type paradigm shows that it is not only fully feasible, but also effective. Definitely, similar (or analogous) bi-partial functions can be defined also for the fuzzy version of the paradigm (FCM). Dvoenko (2014) presents yet another version of the same paradigm ("the meanless k-means"), also transformed using the bi-partial approach.

When the procedure, as exemplified here for two academic instances, is run for more realistic data sets, in terms of their magnitude, the procedure proposed for the approach is to perform the usual k-means algorithms for consecutive decreasing p, starting from a relatively high number (say, $n^{1/2}$, see Owsiński and Mejza, 2007, 2008, Owsiński, 2010), for which the algorithms usually perform very fast, and a lot of repetitions are economised. The results, obtained for the preceding value of p, can be made use of in the next step of the procedure. Behaviour of values of $Q_D{}^S(P)$ is observed and the minimum is registered.

5.3 Some Other Implementations of the Bi-partial Objective Function

5.3.1 Preliminary Remarks—The Objective Function and the Algorithm

We shall now show just a couple of examples for the formulation and interpretation of the bi-partial objective function in cluster analysis. A broader account on the potential formulations and their relation to the existing approaches can be found in Owsiński (1990), as well as in the doctoral dissertation of the present author.

Until now, we have outlined two variants of application of the bi-partial approach in clustering, the one that we refer to as the "leading example", resulting in an MP-like formulation that can be solved approximately via a progressive-merger-like procedure, and an extension to the k-means-type approach, where the algorithmic side was fully taken from the standard k-means, and the bi-partial extension allowed for the control of the algorithm regarding the optimum p, the number of clusters.

In this way we provided the illustration for two levels of application of the bi-partial paradigm. The first, most important level, is namely that of the proper representation of the original clustering problem (the very formulation of the corresponding objective function), and second, the possibility of designing effective algorithms, solving the problem in its bi-partial form. While, as illustrated in several places until now, we can formulate the bi-partial objective function for a wide class of concrete problems, the situation is much more complex with the 'effective algorithms'. We can propose the algorithms similar to that outlined preliminarily in Sect. 5.1.3 for a somewhat limited (even if quite reasonable) class of problems, reflected through appropriate functional representations, but, in general, we do not dispose of a recipe for a 'bi-partial algorithm' that would be at the same time possibly general and convincingly effective.

Yet, of the few additional examples that we present here for clustering, all can be treated with the simile of the progressive merger procedure, as sketched in Sect. 5.1.3. This is, naturally, not to say that all the approaches to clustering could be sensibly formulated according to the bi-partial paradigm, the effective algorithms put apart.

One more remark is due here: in the following sections we present only some selected examples of formulations, leaving quite an important part of the survey of potential formulations and respective algorithms to Chap. 6.

5.3.2 *The Additive Objective Function with a Constant Cluster Cost*

We have, actually, already considered this case in Sect. 4.4, where we analysed the possibility of representing the facility location problem in terms of the explicit bi-partial objective function. Notwithstanding the possible variants of the actual facility location problem, we can represent it in the following manner, for the minimised version of the bi-partial function:

$$Q_D(P) = \sum_q D(A_q); \quad Q^S(P) = p \tag{5.12}$$

which, even if a bit artificial from the clustering perspective, regarding the form of $Q^S(P)$, if appropriately scaled (normalised), can still be treated as a kind of representation of the bi-partial paradigm (at the extreme, as in k-means, even if the objects are definitely most similar to themselves, we suppose that there is an "appropriate" number p of clusters that balances out the tendency towards isolating objects as single-object clusters).

Furthermore, with this kind of bi-partial function, we can also devise a sort of progressive merger procedure, based on the smallest distances between clusters. Actually, for $p = n$ we have, assuming that $D(x) = 0$ for any $x \in E_K$, which is quite natural,

$$Q_D(I) = 0; \quad Q^S(I) = n, \quad \text{i.e.}\, Q(I) = n. \tag{5.13}$$

Then, the question is: can we find a pair of objects, say i^* and j^*, such that

$$D(\{i^*, j^*\}) < 1,$$

so that the value of the overall objective function for the partition, in which only this particular pair is formed, is smaller than in (5.13), i.e. than for the partition $P = I$:

$$D(\{i^*, j^*\}) + n - 1 < n. \tag{5.14}$$

Consequently, it can be concluded that, as long as the value of $D(A_q)$ for a cluster A_q, which is formed through aggregation of any (two) clusters from some preceding partition, is smaller than 1, it pays to proceed with the thus defined aggregation, since $Q(P)$ shall thereby decrease.

In this manner we can easily design in details a progressive merger algorithm that would stop once no longer $D(A_q) < 1$ can be found, meaning that we established a suboptimal solution in terms of $Q(P)$. It suffices for the function $D(A_q)$ to satisfy quite natural and simple conditions (no decrease of the value being possible as we join the objects and clusters that are increasingly more distant) for the thus outlined procedure to determine a good approximation of the optimum solution.

5.3.3 The Case of Minimum Distances and Maximum Proximities

Another reasonable representation of the clustering problem might involve the following, actually quite classical, definitions of distances and proximities for the particular clusters

$$D(A, B) = \min_{i \in A, j \in B} d_{ij}; \quad \text{and} \quad S(A) = \max_{i,j \in A} s_{ij} \tag{5.15}$$

with the following definitions of the bi-partial components of the overall objective function:

$$Q^D(P) = \sum_q \sum_{q' > q} D(A_q, A_{q'}) \quad \text{and} \quad Q_S(P) = \sum_q \text{card}\, A_q \cdot S(A_q), \qquad (5.16)$$

where card A denotes the number of objects in the set A.

This formulation, while offering a relatively well justified model for the clustering problem, allows also for the use of the progressive merger algorithm, very similar to those already proposed, and also leading to a suboptimal solution. The formulation satisfies by itself the conditions for such an algorithm to be applicable, and to stop at the suboptimal solution. We shall yet return to the analysis of the particular case in the subsequent chapter, where the complete precepts of the progressive merger algorithm shall be presented, along with a much broader spectrum of formulation instances.

5.3.4 The Case of Average Distances and Additive Proximities

This is another case, for which it is possible to devise, or rather—simply—use, the progressive merger algorithm, based on direct application of the precepts, announced to be developed in the following chapter (but, actually, entirely analogous to those already presented). In this case, at the level of individual clusters, we refer to the following definitions:

$$D(A, B) = \frac{1}{card\, A \cdot card\, B} \sum_{\substack{i \in A \\ j \in B}} d_{ij}, \quad \text{and} \quad S(A) = \frac{1}{2} \sum_{i,j \in A} s_{ij}, \qquad (5.17)$$

while the respective bi-partial components of the overall objective function, i.e. $Q^D(P)$ and $Q_S(P)$, are simple summations over all clusters, forming the partition.

5.4 Comparing and Assessing Clustering Quality: The Indices and the Principles

5.4.1 Introductory Remarks

We have already stated in Sect. 2.3.8 that there exist virtually dozens of clustering quality indices (and even more of them, if we account for their diverse varieties …). Suffice to note that already in 1985 Milligan and Cooper (1985) referred to 30 "stopping rules" in their study of the progressive merger algorithms and the choice of the number of clusters. These "stopping rules" are, actually, the partition quality indices, meant to aid in the selection of the "proper" partition (and hence also of the number of clusters) in the sequence of those generated by the progressive merger procedures.

It was also pointed out before that these various indices are largely the substitutes, developed in view of the absence of the clustering methods that integrate within one approach the algorithmic side with the appropriate representation of the clustering problem—equivalent to the criterion of goodness of clustering. The bi-partial approach, as presented here, offers, first, a broader perspective on the construction of the respective criteria or objective functions—which can constitute extensions to several of the existing algorithmic paradigms, and, second, under specific conditions, proposes also definite algorithmic procedures, leading to the solution of the clustering problem.

We should also add, against the background of the existing rich literature of this subject,[4] that we do **not** speak here of the so-called *external* quality measures, which refer not only to the set X of data, but also to the (external) information on the "correct" partition of the objects in X (or the known labelling of classes thereof). Thus, we speak solely of the so-called *internal* measures, i.e. the ones that ought to help in comparing partitions or the clustering methods, with which they were obtained, only on the basis of the data set X.

It would be strange indeed, if the existing internal measures had nothing to do with the rationale similar to that behind the bi-partial approach. And, indeed, many, if not most of them can be ultimately seen as representing a sort of bi-partial assessment of partitions of the set I.

If, however, these criteria or quality measures do, in fact, represent the way of thinking that is very much in line with the bi-partial approach, ***what additional value does the bi-partial approach bring, on the top of application of any of those measures?***

1. First, the bi-partial approach proposes a general view of the broadly conceived problem formulation, so that on its basis both various forms of criteria or objective functions can be formulated, and many others can be brought to the bi-partial form.

2. The generality of the approach extends beyond the domain of clustering, as we have seen in the examples here provided. Indeed, in any task from the very broadly conceived area of data analysis, where there is some

[4]In addition to the already mentioned references on the clustering quality measures, i.e. Rendón et al. (2011) and Vendramin, Campello and Hruschka (2010), we would like to note the early paper by Milligan and Cooper (1985), those by Meila (2005), Liu et al. (2010), Zhao and Fränti (2014), Zhao, Xu and Franti (2009), Van Craenendonck and Blockeel (2015), and, first of all, Desgraupes (2013), where 42 indices are collected, as accounted for in an R-package-related project.

"main criterion", usually very intuitive, but also some kind of "altera pars" (e.g. in the case of the number of factors retained in factor analysis), the bi-partial approach can be potentially applied.

3. As it has also been shown—the bi-partial approach is amenable to the design of appropriate algorithms, oriented at the respective criteria as objective functions. There is the main algorithmic avenue, which will be deployed amply in Chap. 6, based on imposing certain conditions on the functions, forming the general bi-partial objective function. Yet, as this has been illustrated for the k-means case, there are also the possibilities to devise workable algorithms, corresponding to the bi-partial approach, for at least some of the existing clustering paradigms, using the algorithmic precepts, proper for these paradigms.

Having said that, in what follows, we shall first recall some of the measures mentioned and analyse them from the perspective of the bi-partial approach, and then forward some remarks, concerning the comparison of algorithms and partitions in general.

5.4.2 *The Exemplary Internal Clustering Quality Measures*

We shall start with the classical measure, proposed initially by Dunn (1974). This measure in its original rendition can be easily transformed so as to be expressed as

$$Q_{D1} = \frac{\min_{q,q'} D(q,q')}{\max_q D(q)} \tag{5.18}$$

with, as usual, q and q' being the indices of clusters, forming the given partition P, $D(.,.)$ being the measure of distance between a pair of clusters, and $D(.)$ being the internal distance measure of a single cluster.

It is also definitely easy to see, especially after expressing the above in terms of logarithms, namely:

$$\log Q_{D1} = \log \min_{q,q'} D(q,q') - \log \max_q D(q) \tag{5.19}$$

that maximising this function is exactly equivalent to maximising the function

$$Q_S{}^D(P) = Q^D(P) + Q_S(P)$$

in which

$$Q^D(P) = \log \min_{q,q'} D(q,q'), \text{ and } Q_S(P) = M - \log \max_q D(q),$$

with M being an appropriately big number.

There exists quite a variety of criteria, broadly analogous to Q_{D1}, as well as a number of the concrete implementations of the basic form (5.18). One of the variants, meant for the fuzzy clustering methods, was proposed by Bezdek et al. (1999):

$$Q_{D2} = \frac{\min_{q,q'} \dfrac{\sum_{i=1}^{n}\sum_{j=1}^{n} u_{iq}u_{jq'}d(i,j)}{\left(\sum_{i=1}^{n} u_{iq}\right)\left(\sum_{i=1}^{n} u_{iq'}\right)}}{\max_q 2\dfrac{\sum_{i=1}^{n} u_{iq}d(i,q)}{\sum_{i=1}^{n} u_{iq}}} \tag{5.20}$$

where u_{iq} denote the memberships, $\in[0, 1]$, of objects, indexed i, in fuzzy clusters, indexed q. It can easily be seen that (5.20) is, in fact, just a variety of (5.18).

Quite an analogous reasoning can be applied to the similarly popular Xie-Beni index (Xie and Beni, 1991), designed initially also for the fuzzy clusters and partitions, but which can be equally well applied, in an appropriate manner, to the crisp partitions. This minimised index is equivalent to the following form:

$$Q_{XB} = \frac{\frac{1}{n}\sum_{q=1}^{p}\sum_{i=1}^{n} u_{iq}^{a} d^2(i,q)}{\min_{q,q'} d^2(q,q')}. \tag{5.21}$$

This form, again, as this was done with respect to (5.18), can be easily turned into the bi-partial function, the corresponding variant of this function being the minimised "dual":

$$Q_D{}^S(P) = Q_D(P) + Q^S(P)$$

in which the component $Q_D(P)$ is expressed through the nominator in (5.21), while $Q^S(P)$ is expressed by the denominator, thereby establishing the correspondence.

Another pertinent example is provided by the criterion, used by Stanfel (1983), namely

$$Q_{S1}(P) = \frac{\sum_{q=1}^{p}\sum_{i\in A_q}\sum_{j\in A_q} d_{ij}}{\sum_{q=1}^{p} card\, A_q(card\, A_q - 1)} - r\frac{\sum_{q=1}^{p-1}\sum_{q'=q+1}^{p}\sum_{i\in A_q}\sum_{j\in A_{q'}} d_{ij}}{\sum_{q=1}^{p-1}\sum_{q'=q+1}^{p} card\, A_q card\, A_{q'}}. \tag{5.22}$$

In this criterion r is a predefined coefficient, $r\in[0,1]$, the "method parameter". The criterion (5.22) is minimised, of course, and it is not difficult to show its conformity with the bi-partial objective function $Q_D{}^S(P)$.

One of the most popular criteria of partition goodness is the one, proposed by Calinski and Harabasz (1974). It refers, on the one hand, to the commonly used variance ("sum of squares") framework, and on the other—it is, in fact, a "stopping rule" for the progressive merger procedures, as studied by Milligan and Cooper (1985). We shall express it here in a more general form, closer to our line of reasoning, namely:

$$Q_{CH}(P) = \frac{\frac{1}{p-1}\sum_{q=1}^{p} n_q d^2(x^q, x^{av})}{\frac{1}{n-p}\sum_{i=1}^{n} d^2(x_i, x^q)}, \tag{5.23}$$

where, as before, n_q is the number of objects in cluster q, x^q denotes here the average for the cluster q, and x^{av}—the average for the entire set X.

In the original statement the sum in the nominator is the trace of the inter-cluster covariance matrix ("SSB"—between cluster sum of squares), and the sum in the denominator—the trace of the intra-cluster covariance matrix ("SSW"—within cluster sum of squares). In any case, it is easy to see that this maximised criterion can be directly expressed in terms of the bi-partial objective function.[5]

We shall yet return in this book to the issue of analogy between the bi-partial objective function and the criteria or approaches, proposed in the pertinent rich literature, first of all in the subsequent section, and then in some other places. Now, however, we would like to note that some of proposed criteria, apparently following the same kind of reasoning, cannot be easily "arithmetically" turned into the bi-partial form, as introduced here. An example to the point is provided by another popular criterion, proposed by Davies and Bouldin (1979), namely:

$$Q_{DB} = \frac{1}{p}\sum_{q=1}^{p} \max_{q' \neq q} \frac{D'(q) + D'(q')}{D(q, q')}. \tag{5.24}$$

The problem here is, of course, in the change of the ordering of the "levels of perception" (individual clusters vs. the overall partition), which involves a non-linear transformation. At the same time, even though we cannot transform (5.24) so as to obtain a simile of the bi-partial objective function, it is obvious that (5.24), which is minimised, of course, represents, in general terms, the same perspective as the bi-partial approach.

[5]The sum of within-cluster and between-cluster sums of squares is constant—the total sum of squares. For this reason it is sometimes asked why bother with accounting for both of these, if maximisation of one leads to minimisation of the other. The fact that the here quoted indices account for these two components (inter- and intra-cluster), in their various renditions, demonstrates very pungently the reason behind, and so also the validity of the bi-partial approach. The logic of selecting the smaller distances ("squares") for inclusion in the clusters and bigger for separation of clusters is exactly the same as in our leading example of the bi-partial objective function.

5.4.3 The Founding Ideas

It should be noted that there are some references, which seem to be of special importance for the origins of the bi-partial approach. We shall recall them in this subsection.

Thus, Regnier (1965) postulated the search for the "central partition" among all partitions of the set of data objects. Regnier analyses the space of partitions, E_P, as the set of vertices of the multidimensional cube, these vertices being described by the numbers $y_{ij} \in \{0,1\}$, where $y_{ij} = 1$ when objects i and j belong to the same cluster, while $y_{ij} = 0$ in the opposite case (or, actually, vice versa, depending on whether we consider distances or proximities, see below). When, in this context, the distances d_{ij} or proximities s_{ij} are normalised, due to division by d^{max} or, respectively, s^{max}, then the resulting values of d^*_{ij} or s^*_{ij} are represented by the points, situated in $R^{n \times n}$[6] inside the cube, mentioned before, representing the partitions through its vertices. Now, the task of cluster analysis would consist in finding this vertex in E_P, which is located the closest to the set of points, corresponding to d^*_{ij} or s^*_{ij}, this vertex being equivalent to the sought central partition.

Regnier proposed to apply an exceptionally simple criterion for this task. This (maximised) criterion, after a simple transformation, can be expressed as

$$Q_{R1} = \sum_{q=1}^{p} \sum_{i,j \in A_q} (s^*_{ij} - 0.5) \tag{5.25}$$

(see the simple illustrations in Figs. 5.3 and 5.4) or, quite equivalently, but closer to our framework:

$$Q_{R2} = \sum_{q=1}^{p} \sum_{i,j \in A_q} (s^*_{ij} - 0.5) - \sum_{q=1}^{p-1} \sum_{q'=q+1}^{p} \sum_{i \in A_q} \sum_{j \in A_{q'}} (s^*_{ij} - 0.5). \tag{5.26}$$

It can easily be seen that (5.26) brings Regnier's criterion very close, indeed, to the bi-partial formulation. On the other hand, the algorithms, proposed by Regnier for solving the thus formulated problem turned out to be quite ineffective.

It seems that quite independently of Regnier two authors, Fortier and Solomon (1966), published just a bit later a proposal for the minimised clustering criterion, which can be expressed as

$$Q_{FS} = \sum_{q=1}^{p} \sum_{i,j \in A_q} (d^*_{ij} - r) - \sum_{q=1}^{p-1} \sum_{q'=q+1}^{p} \sum_{i \in A_q} \sum_{j \in A_{q'}} (d^*_{ij} - r) \tag{5.27}$$

and includes a parameter $r \in [0, 1]$.

---

[6] Actually, due to the assumed symmetricity of distances and similarities, the dimensionality of this space is $1/2(n(n - 1))$, see the illustrations further on.

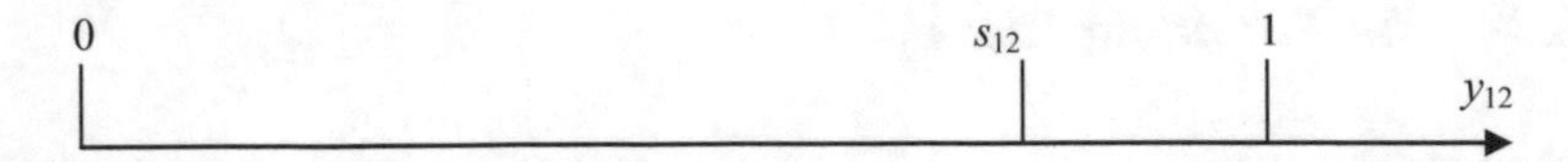

Fig. 5.3 An example regarding Regnier's idea for $n = 2$, i.e. the unidimensional case; the value of proximity s_{12} indicates that joining of the two objects is more advisable than separating them (abstraction is made here of the normalisation issue)

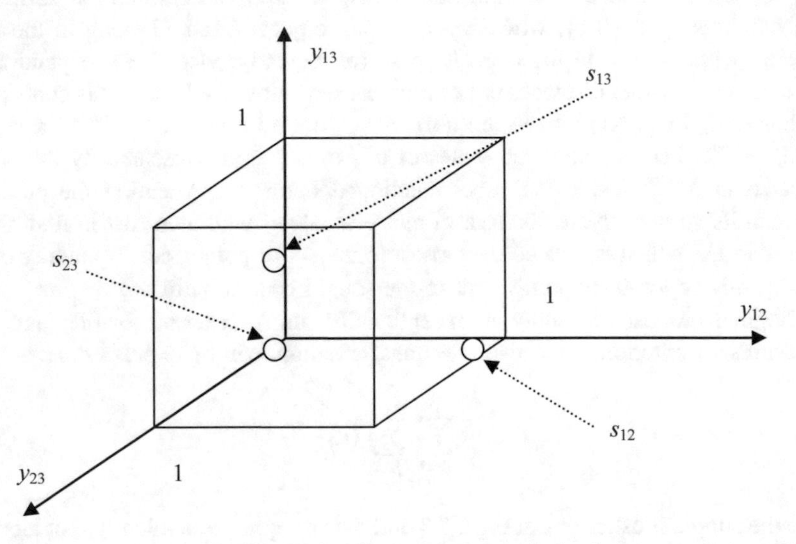

Fig. 5.4 An example regarding Regnier's idea for $n = 3$; intuitively, the proper solution appears to be {{1, 2}, 3}

Besides the similarity with the bi-partial form, the introduction of the parameter r is significant, as we shall see later on in the development of the algorithm, related to the bi-partial approach. Yet, Fortier and Solomon did not use, actually, this parameter in any effective manner, analysing only the case of $r = 1/2$, making the function equivalent to (5.26). Likewise, the algorithms proposed were not effective, so that the sole example considered in the article concerned $n = 19$ objects.

Another year later, Rubin (1967) published a broader article, in which he postulated the properties of the clustering methods and proposed also an objective function, one of whose forms can be brought to the following expression, which is maximised:

$$Q_{R1} = \frac{1}{n}\left(r\sum_{q=1}^{p}\sum_{i,j\in A_q} s_{ij}^{*} - (1-r)\sum_{q=1}^{p}\sum_{i\in A_q}\sum_{j\in A_{q^*(i)}} s_{ij}^{*} \right). \tag{5.28}$$

Here, $A_{q^*(i)}$ is the cluster, for which

$$\sum_{j \in A_{q^*(i)}} s^*_{ij} = \max_{\{q | i \notin A_q\}} \sum_{j \in A_q} s^*_{ij}. \tag{5.29}$$

Due to the complexity of this proposal, Rubin has not even suggested any algorithm for optimising the partitions with this function. On the other hand, he analysed, although quite superficially, the effects of change in the value of r over the interval [0, 1], but has not drawn from this analysis any constructive conclusions.

Then, three years later, Ducimetière (1970) published a survey paper, in which, following mainly Rubin (1967) he formulated the objective function as follows:

$$Q_{DP} = \sum_{q=1}^{p} \sum_{i,j \in A_q} (s^*_{ij} - r) \tag{5.30}$$

the parameter r having the same character as before, and function (5.30) being, of course, maximised.

Of essential importance is the fact that Ducimetière mentions the possibility of constructing for (5.30) a suboptimising algorithm of progressive merger character, for the parameter r decreasing from 1 to 0. Such an algorithm would be identical with the classical average linkage algorithm (see, e.g., Lance and Williams, 1966, 1967). Yet, Ducimetière does not elaborate on this any further.

Next, after a couple of years, in 1977 de Falgucrolles (1977) partly summarised the developments in the here outlined direction (without mentioning, anyway, the work by Fortier and Solomon). De Falguerolles formulated in this article the following objective function:

$$Q_{DF} = r \sum_{q=1}^{p} \sum_{i,j \in A_q} s^*{}_{ij} + (1 - r) \sum_{q=1}^{p-1} \sum_{q'=q+1}^{p} \sum_{i \in A_q} \sum_{j \in A_{q'}} d^*{}_{ij} \tag{5.31}$$

where he assumed that $s^*{}_{ij} = 1 - d^*{}_{ij}$. De Falguerolles analysed in detail the properties of (5.31) and provided its alternative formulations, but, with respect to the algorithmic suggestions, did not even develop over Ducimetière, proposing merely the use of exchange procedures or k-means similes.

We shall close this section with the mention of the work by Marcotorchino and Michaud (1979, 1982), already referred to in Sect. 5.1.2. As indicated there, they formulated a mathematical programming problem, which we repeat here for completeness:

$$\text{maximise} \sum_{i,j \in I} \left(y_{ij} s_{ij} + (1 - y_{ij}) d_{ij} \right) \tag{5.1}$$

where $y_{ij} = 1$ when objects i and j belong to the same cluster, and $y_{ij} = 0$ otherwise, with the following additional constraints:

$$
\begin{aligned}
&y_{ij} \in \{0, 1\}, \quad \forall i, j, \\
&y_{ij} = y_{ji}, \quad \forall i, j, \text{meaning symmetry} \\
&y_{ij} + y_{jv} - y_{iv} \leq 1, \quad \forall i, j, v, \text{meaning transitivity.}
\end{aligned} \tag{5.2}
$$

The very form of the objective function was in principle the same as developed in the here reported line of research, but the proposal was to treat the problem as a mathematical programming task, to be solved with powerful computing resources. The technical aspect was somewhat alleviated by the establishment that the solutions to this problem, (5.1)–(5.2), are always binary, even if y_{ij} are allowed to be continuous, i.e. $\in$ [0, 1]. Yet, it is obvious that for the current practical situations, in which one often deals with very high numbers of objects, this way of proceeding may turn out impracticable (the number of the transitivity constraints being at the order of n^3).

5.4.4 Assessing Clustering Quality

We shall now return to the basic question of this part of the book, namely—how to assess the quality of an algorithm and of its results? We would like to emphasise again that we speak of the "internal" criteria, that is—we assume there are no "training examples" nor "benchmarks", the latter at least in the "absolute" sense, regarding the vast majority of the data sets, used for this purpose.[7]

In this short section we do not intend to display any "theory of clustering evaluation", the more so since we are of the opinion that—as indicated already in this volume—there may exist various justified convictions as to what the "correct clustering" of a given set of objects is, especially when the given problem is not expressed in a manner adequate for the clustering task in general, like this is done in the bi-partial approach. The often, but not always, repeated postulate of finding "natural" groups in data can at most lead to the discussion what this "naturalness" is. On the other hand, there may also exist a need of partitioning a set of objects into subsets that is driven by some pragmatic interest, expressed, say, in sheer monetary terms, like in the facility location problems, where "naturalness" has indeed very little to say.

Yet, there are, definitely, some quite common sense premises, which can serve at least the purposes of basic assessment or even "elimination". Some of the principles thereof were provided in Owsiński (2003, 2004) in terms of "ideal structures", which can serve as references for the assessment of clustering methods.

[7]It must be kept in mind that the assignment to classes in various "benchmark" data sets has very different statuses, both in terms of "certainty" of assignment and of the relation between the data provided and the assignment (e.g. the question: can there exist any method that would classify entirely correctly the given set without additional information?).

There seem to be two most general kinds of such references, namely

(1) the "fully random" or "(quasi)equidistant" reference, and
(2) the "ideal clustering structure" reference,

each of these (but especially the latter) with a variety of possible configurations.

While it is usually postulated with respect to case (1) that the methods ought not indicate any partitioning, the present author is of the opinion—especially with respect to the "haphazard", but close to "equidistant", rather than "strictly equidistant" distribution of objects—that a method ought to yield some partitioning, preferably depending upon a parameter, with the indication of very feeble distinction of the "quality" of the potential partitions (see Fig. 5.5).

In situations like that in Fig. 5.5 it would be advisable to have the method showing that along a series of partitions, ranging from $P^1 = I$ (all the objects separately) to some $P^p = \{I\}$ (all the objects in one cluster) there are only very limited, marginal changes of partition quality.

An instance of data that could be used as type (2) reference is provided in Fig. 5.6. In this kind of setting one would expect the method to find five clusters (potentially—perhaps four, groups 3 and 5 being put together), but also to quantify the differences in quality between the solutions, indicating various numbers of clusters.

It might not only be interesting, but also instructive to see (and assess) the differences in quality between the solution into the five groups shown in the figure, and, say, the partition into {1}, {2}, {3, 4, 5}, and, ultimately, {I} = {1, 2, 3, 4, 5}.

Regarding this sort of data sets, the reference Owsiński (2003, 2004) provides a number of definitions of "ideal structures", starting with the obviously the strongest one, when

$$d_P^{\max} = \max_q d(i,j|i,j \in A_q) < d_{PP}{}^{\min} = \min_{q \neq q'} d(i,j|i \in A_q, j \in A_{q'}),$$

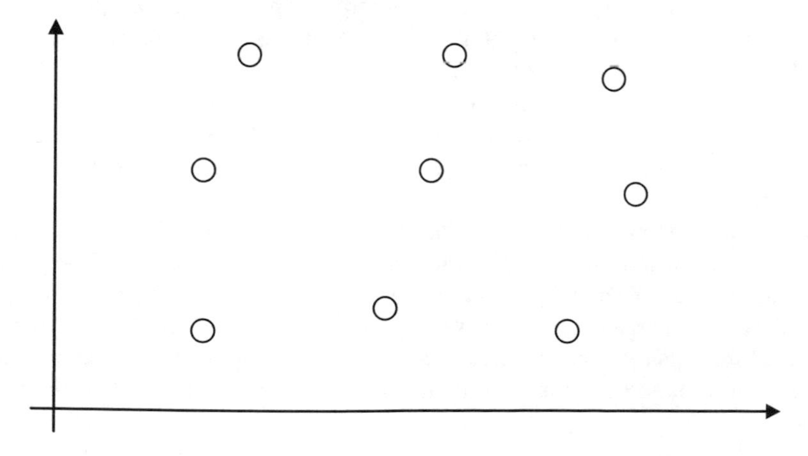

Fig. 5.5 An example of data for case (1) of the general references

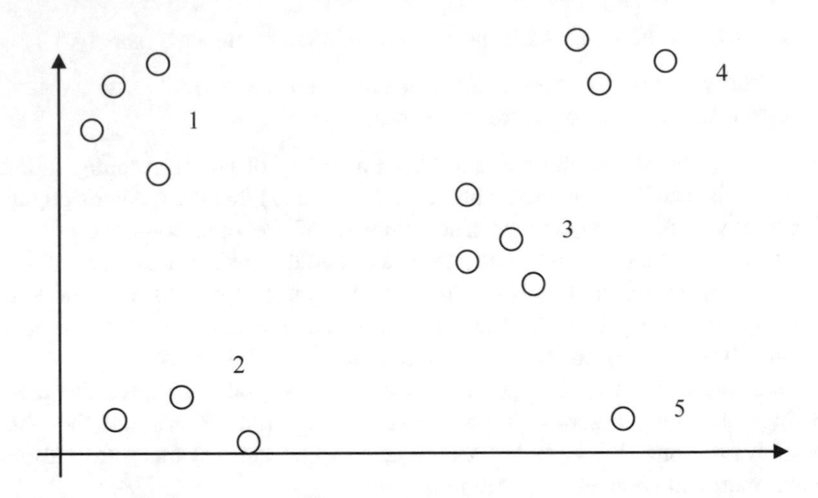

Fig. 5.6 An example of data for case (2) of the general references

and it is expected that the ("correct") clustering methods distinctly point out to the existence of such a structure. The reference quoted then moves in the direction of the less restrictive structure conditions, where the indications must not be that strict any more. These consecutive structures could be effectively used to check the properties of the clustering methods.

5.5 Summarising the Place of the Bi-partial Approach and the Algorithms Thereof

Having presented several instances of the problems in data analysis, and then more specifically in clustering, for which the bi-partial approach can be applied, along with some hints as to the potential realisation of the respective algorithms, we would like to give a glimpse of a very simplified picture, showing the place of the paradigm, and the related algorithms, against the domains here considered.

Thus, Fig. 5.7 displays the sketchy image of the place, occupied by the bi-partial paradigm, and the algorithms that can be devised in the framework of this paradigm.

It is quite obvious that the bi-partial paradigm can be applied to only a definite subset of the concrete—even if general—problems from data analysis. The primary domain, in which this paradigm finds an obvious application, is cluster analysis, but even there an important part of the existing approaches cannot be transformed to the bi-partial formulation. The setups, for which effective algorithms can be proposed, form an even narrower class of formulations, even though, as we shall see in the

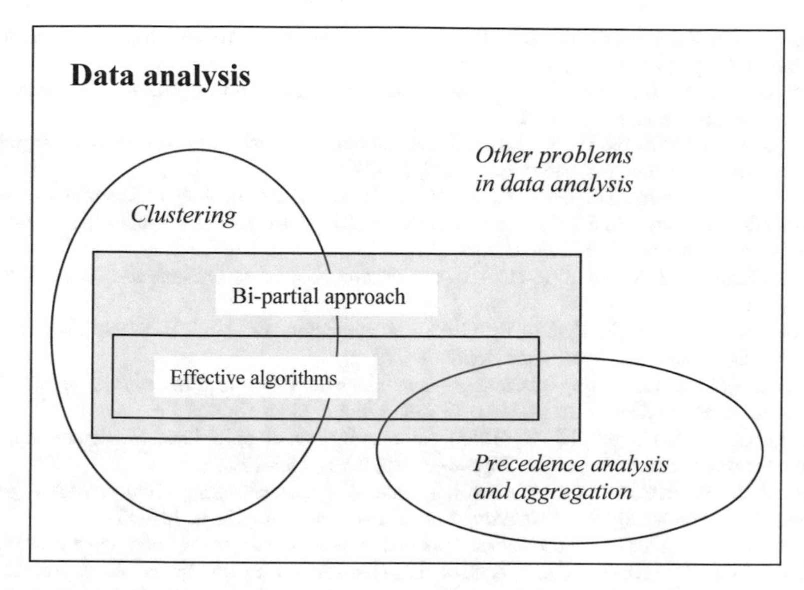

Fig. 5.7 Schematic view of the place of bi-partial approach and the potential effective algorithms against the background of the domain of data analysis

next chapter, this class is definitely not negligible against the background of the entire domain of cluster analysis. This schematic illustration shall yet be commented upon in further parts of the volume.

References

Bezdek, J. C., Keller, J., Krisnapuram, R., & Pal, N. R. (1999). Fuzzy models and algorithms for pattern recognition and image processing. In *The handbooks of fuzzy sets* (Vol. 4), Springer.

Calinski, T., & Harabasz, J. (1974). A dendrite method for cluster analysis. *Communications in Statistics, 3,* 1–27.

Davies, D. L., & Bouldin, D. W. (1979). A cluster separation measure. *IEEE Transactions on Pattern Analysis and Machine Intelligence, 1*(2), 224–227.

de Falguerolles, A. (1977). Classification automatique: Un critère et des algorithms d'échange. In E. Diday & Y. Lechevallier (Eds.), *Classification automatique et perception par ordinateur*. Le Chesnay: IRIA.

Desgraupes, B. (2013). *Clustering indices.* CRAN-R-Project. https://cran.r-project.org/web/packages/clusterCrit/…/clusterCrit.pdf.

Ducimetière, P. (1970). Les méthodes de la classification numérique. *Revue de Statistique Appliquée, 18*(4), 5–25.

Dunn, J. (1974). Well separated clusters and optimal fuzzy partitions. *Journal of Cybernetics, 4*(1), 95–104.

Dvoenko, S. (2014). Meanless *k*-means as *k*-meanless clustering with the bi-partial approach. In *PRIP'2014. Proceedings of 12th International Conference on Pattern Recognition and Image Processing* (pp. 50–54). Minsk, Belarus: UIIP NASB, May 28–30, 2014.

Fortier, J. J., & Solomon, H. (1966). Clustering procedures. In P. Krishnaiah (Ed.), *Multivariate analysis I* (pp. 493–506). London: Academic Press.

Lance, G. N., & Williams, W. T. (1966). A generalized sorting strategy for computer classifications. *Nature, 212,* 218.

Lance, G. N., & Williams, W. T. (1967). A general theory of classification sorting strategies. 1. Hierarchical systems. *Computer Journal, 9,* 373–380.

Liu, Y., Li, Z., Xiong, H., Gao, X., & Wu, J. (2010) Understanding of internal clustering validation measures. In *2010 IEEE International Conference on Data Mining* (pp. 911–916). IEEE, https://doi.org/10.1109/icdm2010.35.

Marcotorchino, F., & Michaud, P. (1979). *Optimisation en analyse ordinale des données*. Paris: Masson.

Marcotorchino, F., & Michaud, P. (1982). Aggrégation de similarités en classification automatique. *Revue de Statistique Appliquée, 30,* 2.

Meila, M. (2005). Comparing clusterings—an axiomatic view. In *Proceedings of the 22nd International Conference on Machine Learning.* Bonn, Germany.

Milligan, G. W. & Cooper, M. C. (1985) An examination of procedures for determining the number of clusters in a data set. *Psychometrika*, 50(2), 159–179.

Owsiński, J. W. (1990). On a new naturally indexed quick clustering method with a global objective function. *Applied Stochastic Models and Data Analysis, 6,* 157–171.

Owsiński J. W. (2003). Group choice: opinion structure, consensus, and cluster analysis. Taksonomia 10. Klasyfikacja i analiza danych—teoria i zastosowania, K. Jajuga & M. Walesiak, eds. Prace Naukowe AE we Wrocławiu, no. 988, Wyd. AE im. Oskara Langego we Wrocławiu, Wrocław, 332–342.

Owsiński J. W. (2004). Group opinion structure: The ideal structures, their relevance and effective use. In: D. Baier & K.-D. Wernecke, (Eds.), *Innovations in Classification, Data Science, and Information Systems. Proceedings of 27th Annual GfKl Conference, University of Cottbus*, (pp. 471–481), Heidelberg-Berlin: Springer-Verlag, 12–14 March, 2003.

Owsiński, J. W. (2010). On a two-stage clustering procedure and the choice of objective function. In *Computer Data Analysis and Modeling: Complex Stochastic Data and Systems. Proceedings of the 9th International Conference* (pp. 157–165). Minsk, September 7–11, 2010. Publishing center of BSU, Minsk.

Owsiński, J. W. (2011) The bi-partial approach in clustering and ordering: The model and the algorithms. In *Statistica & Applicazioni*. Special Issue (pp. 43–59).

Owsiński, J. W. (2012) Clustering and ordering via the bi-partial approach: The rationale, the model and some algorithmic considerations. In J. Pociecha & R. Decker (Eds.), *Data analysis methods and its applications* (pp. 109–124). Warszawa: Wydawnictwo C. H. Beck.

Owsiński, J. W., & Mejza, M. T. (2007). On a new hybrid clustering method for general purpose use and pattern recognition. In *Proceedings of the International Multiconference on Computer Science and Information Technology* (Vol. 2, pp. 121–126). http://www.papers2007.imcsit.org/.

Owsiński J. W., & Mejza M. T. (2008). A new hybrid clustering method: from "subassemblies" to "shapes". In O. Hryniewicz, A. Straszak, J. Studziński, (Eds.), Badania operacyjne i systemowe: środowisko naturalne, przestrzeń, optymalizacja. Badania Systemowe, 63. Instytut Badań Systemowych PAN, Warszawa, pp. 341–348.

Regnier, S. (1965). Sur quelques aspects mathématiques des problems de classification automatique. *ICC Bulletin, 4,* 175–191.

Rendón, E., Abundez, I., Arizmendi, A., & Quiroz, E. M. (2011). Internal versus external cluster validation indexes. *International Journal of Computers, Communications & Control (IJCCC), 5*(1), 27–34.

Rubin, J. (1967). Optimal classification into groups: An approach for solving the taxonomy problem. *Journal of Theoretical Biology, 1,* 103–144.

Stanfel, L. E. (1983). Applications of clustering to information system design. *Information Processing & Management, 19*(1), 37–50.

Van Craenendonck, T., & Blockeel, H. (2015). Using internal validity measures to compare clustering algorithms. In *Poster from Benelearn Conference*, 2015, https://lirias.kuleuven.be/handle/123456789/504705.

Vendramin, L., Campello, R. J. G. B., Hruschka, E. R. (2010). Relative clustering validity criteria: A comparative overview. *Wiley InterScience* https://doi.org/10.1002/sam.10080.

Xie, X. L., & Beni, G. (1991). A validity measure for fuzzy clustering. *IEEE Transaction on Pattern Analysis and Machine Intelligence, 13*(8), 841–847.

Zhao Q., Xu M., & Fränti, P. (2009). Sum-of-squares based cluster validity index and significance analysis. In M. Kolehmainen et al. (Eds.), *ICANNGA 2009. LNCS 5495* (pp. 313–322). Springer Verlag.

Zhao, Q., & Fränti, P. (2014). WB-index: A sum-of-squares based index for cluster validity. *Data & Knowledge Engineering, 92,* 77–89.

Chapter 6
The General Sub-optimisation Algorithm and Its Implementations

In this chapter we shall present more amply the principles of construction and the properties of the subpotimisation algorithm that is associated with the bi-partial paradigm, and which can be designed and realised for at least some of its actual implementations. We have already presented the possibility of taking advantage of the general precepts of bi-partial paradigm in devising some kind of algorithm, like in the case of k-means, but the principles here introduced shall be more directly connected with the concrete bi-partial objective functions. Imposing the conditions, which allow for the construction of the algorithm means, of course, narrowing down the range of objective function implementations, but there is always a price to pay, although in this case several of the already signalled implementations of the bi-partial objective function fulfil these conditions.

6.1 Basic Properties of the Objective Function

The basic conditions that we require of the components of the objective function in order for the proposed algorithm to work are as follows (note that these conditions are formulated here for the maximised function $Q_S^D(P) = Q_S(P) + Q^D(P)$), for $Q^D(P)$:

$$Q^D(P) \le Q^D\big(P^H(q',q'')\big), \text{ that is, } \quad Q^D(P_H(q^*,q^{**})) \le Q^D(P) \\ \forall q', q'', q^*, q^{**} \in K_P = \{1, 2, \ldots, p_P\} \tag{6.1}$$

and for $Q_S(P)$:

$$Q_S(P) \ge Q_S\big(P^H(q',q'')\big), \text{ that is, } \quad Q_S(P_H(q^*,q^{**})) \ge Q_S(P) \\ \forall q', q'', q^*, q^{**} \in K_P = \{1, 2, \ldots, p_P\} \tag{6.2}$$

J. W. Owsiński, *Data Analysis in Bi-partial Perspective: Clustering and Beyond*, Studies in Computational Intelligence 818,
https://doi.org/10.1007/978-3-030-13389-4_6

where $P^H(q',q'')$ is the partition, formed out of P through splitting of some cluster $A_q \in P$ into clusters $A_{q'}$ and $A_{q''}P^H(q',q'')$, while $P_H(q^*,q^{**})$ is the partition, formed out of partition P through the merger of clusters A_{q*} and $A_{q**} \in P$.

It will be further assumed that the equalities take place in (6.1) when

$$D(A_{q'},A_{q''}) = 0 \quad \text{or} \quad D(A_{q*},A_{q**}) = 0 \tag{6.3}$$

and that

$Q^D(P)$ and $Q_S(P)$ are non-decreasing functions of their components, that is, respectively,

$$D(A_q,A_{q'}) \text{ and } S(A_q). \tag{6.4}$$

The last of the here introduced more detailed conditions states that

$$Q^D(P) - Q^D(P_H(q,q')) \text{ is a non-decreasing function of } D(A_q,A_{q'}). \tag{6.5}$$

Assuming the above conditions fulfilled, we can formulate now the following two properties:

Property 1. *When conditions* (6.1–6.3) *are fulfilled, then the two inequalities below hold true*:

$$\max_{P\in E_p} Q^D(P) \leq \max_{P\in E_{p+1}} Q^D(P) \tag{6.6}$$

and

$$\max_{P\in E_p} Q_S(P) \geq \max_{P\in E_{p+1}} Q_S(P), \tag{6.7}$$

where $E_p=\{P\colon cardP = p\}$, *and the equalities in* (6.6), (6.7) *take place for* (6.3).

Property 2. *When conditions* (6.1–6.3) *are fulfilled, then the two equalities below hold true*

$$\max_{P\in E_P} Q^D(P) = Q^D(I) \tag{6.8}$$

and

$$\max_{P\in E_P} Q_S(P) = Q_S(\{I\}), \tag{6.9}$$

where, as usual, E_P *denotes the set of all partitions, card* $I = n$, *and card* $\{I\} = 1$.

Property 1 is a simple consequence of the assumptions adopted here, and we do not show the proof here for shortness, while Property 2 results directly from Property 1.

It can be stated that the conditions and properties, here introduced, add up to the general property of the global components of the objective function, which can be

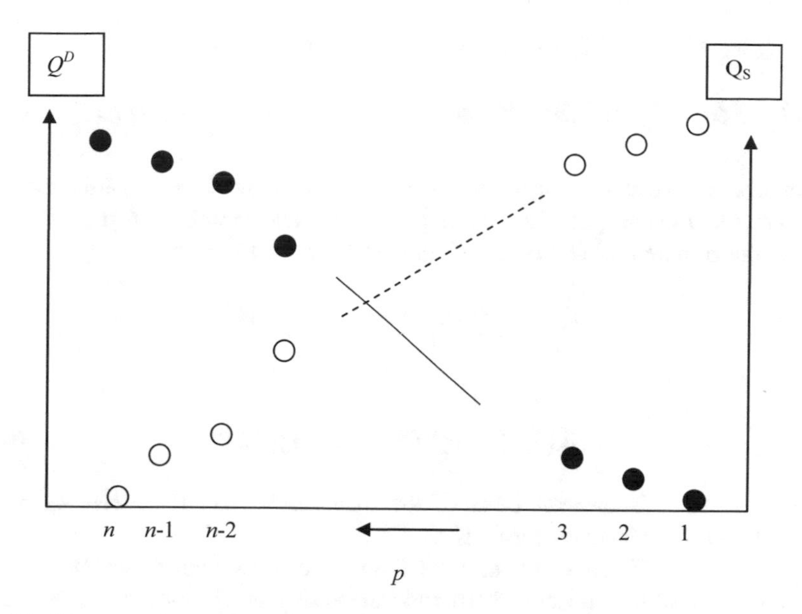

Fig. 6.1 Schematic view of the proposed "opposite monotonicity" of the global components of the bi-partial objective function. The points indicate the maxima over all partitions for consecutive values of p

referred to as "the opposite monotonicity with respect to p". A schematic illustration for the course of values of these components, conform to the properties here introduced, is provided in Fig. 6.1.

The assumptions postulated are not fully in agreement with intuition, especially if we consider that it is at the level of the global components of the objective function, say $Q^D(P)$ and $Q_S(P)$ i.e. those we are analysing here, that the ultimate measurement of the quality of partition is being made. In the bi-partial approach, though, we explicitly propose that such an ultimate measurement can only be made at the level of the entire ("global") objective function, where the two parts meet. This fact allows for a kind of compensation of the apparent lack of intuitive appeal at the lower level of perception.

Let us also add that, obviously, the entirely analogous properties can be introduced for the dual function $Q_D^S(P)$, which is minimised.

6.2 The General Sub-optimisation Procedure

6.2.1 The Algorithmic Form of the Objective Function

From now on, we shall be using the parameterised form of the bi-partial objective function, without changing the essential notations. The parametric forms of the dual pair of the bi-partial global objective functions are as follows:

$$Q_S^D(P,r) = rQ^D(P) + (1-r)Q_S(P) \tag{6.10}$$

and

$$Q_D^S(P,r) = rQ^S(P) + (1-r)Q_D(P) \tag{6.11}$$

where $r \in [0,1]$ is the parameter of the method. We shall proceed, as before, primarily with the analysis for (6.10).

The forms (6.10) and (6.11) have been introduced for algorithmic purposes, but it is obvious that maximising (6.10) and minimising (6.11) for $r = 1/2$ over partitions is equivalent, respectively, to maximising and minimising the original forms of the function.

Let us first note that, according to Property 2, we have

$$\max_{P \in E_P} Q_S^D(P,1) = \max_{P \in E_P} Q^D(P) = Q^D(I) \tag{6.12}$$

and

$$\max_{P \in E_P} Q_S^D(P,0) = \max_{P \in E_P} Q_S(P) = Q_S(\{I\}). \tag{6.13}$$

Now, let us observe that for a definite P, and hence for definite values of $Q^D(P)$ and $Q_S(P)$, the expression (6.10), establishing the value of $Q_S^D(P,r)$, defines a segment of the straight line with respect to the variable r, taking values from [0,1]. Hence, for all $P \in E_P$, one obtains a family of straight lines, differing by their slope with respect to the axis Or of the variable r and by the value at the point of crossing the axis OQ, showing the values of the objective function.

From (6.12) and (6.13) one can easily draw the conclusion that the steepest (positive) slope is achieved by the segment of the line, defined by the expression

$$rD(I) - (1-r) \cdot 0 = rD(I), \tag{6.14}$$

while the most downward slope is achieved by the segment of the line, defined by the expression

$$r \cdot 0 + (1 - r)S(I) = S(I) - r \cdot S(I). \tag{6.15}$$

Thereby, we have defined a certain function of the variable r, namely

$$Q_S^D(r) = \max_{P \in E_P} \{Q_S(P) + r(Q^D(P) - Q_S(P)\}. \tag{6.16}$$

This function is represented by a broken line, composed of the fragments of the segments of the straight line, mentioned already above. These of the segments are included in the function, which attain the highest value for a given r. In conformity with the remark, forwarded before, we have, of course, from (6.16):

$$2 \cdot Q_S^D(0.5) = \max_{P \in E_P} Q_S^D(P). \tag{6.17}$$

It is obvious that the characteristics of the function $Q_S^D(r)$ are of crucial interest for us. We can formulate already now two further properties of this function.

Property 3. *The function $Q_S^D(r)$ is a broken line, composed of a finite number of non-zero length segments.*

To show that Property 3 is true, let us first note that $Q_S^D(r)$ is a broken line by definition, as the maximum of a family of straight lines. Besides, E_P is a finite set, and so $Q_S^D(r)$ is composed of a finite number of segments. The non-zero length of the segments results from the following reasoning:

Assume there exists a P^*, which maximises $Q_S^D(P,r)$ only for a single value of r, namely for r^*. There exists a right-hand neighbourhood of the point r^*, within which a similar situation does not occur. Thus, there exists an interval (r^*, r^{**}), in which the function $Q_S^D(P, r)$ is maximised by a certain partition P^{**}, different, of course, from P^*. From the continuity of the linear function it follows that the line $Q_S^D(P^{**}, r)$ includes also the point $(Q_S^D(P^*, r^*), r^*)$. Were it not so, there would exist a certain right-hand neighbourhood of r^*, for which P^* would remain the maximising partition. Thus, since

$$Q_S^D(P^{**}, r^*) = Q_S^D(P^*, r^*),$$

meaning that P^{**} maximises, as well, the function $Q_S^D(P,r)$ at point r^*, we can assume, by definition, that from the set of partitions, constituting the basis for establishment of the function $Q_S^D(r)$, we eliminate the ones that correspond to just single points in r and replace them by the ones that correspond to, say, the right-hand neighbourhood of such values of r. Analogous reasoning can be carried out for the left-hand neighbourhood, as well as for multiple partitions, maximising simultaneously, but uniquely at some point r^*.

Property 4. *The function $Q_S^D(r)$ is convex for $r \in [0,1]$.*

Consider, in this context, an interval $[r^*,r^{**}]$, over which the function $Q_S^D(r)$ corresponds to some partition P^*. It can be concluded from the previous considerations that this interval can be treated as closed. The slope of the function $Q_S^D(r)$

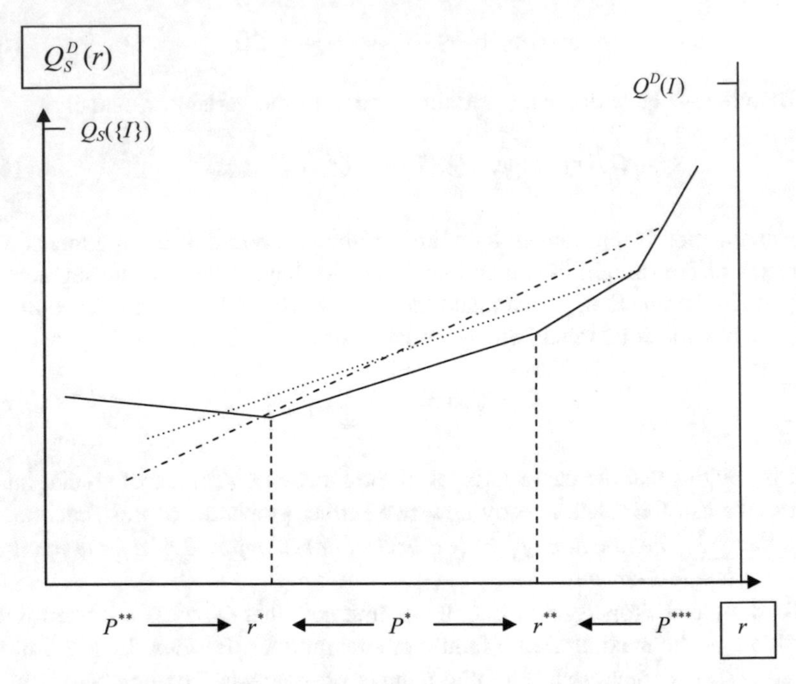

Fig. 6.2 Illustration for the convexity of the function $Q_S^D(r)Q^D(P^{**}) - Q_S(P^{**}) > Q^D(P^*) - Q_S(P^*)$: $- - - - Q^D(P^*) - Q_S(P^*) > Q^D(P^{***}) - Q_S(P^{***})$: ······

over this interval remains constant and equal $Q^D(P^*) - Q_S(P^*)$. Property 4 states that the slope on the segment immediately neighbouring on the left is smaller, while on the segment to the right it is bigger. Were it not so, then it would not be P^* that will maximise $Q_S^D(r)$ over $[r^*,r^{**}]$, but either the partition, corresponding to the segment to the left, or the one to the right. (A schematic illustration of this reasoning is provided in Fig. 6.2.)

So, for $r < r^*$, in the immediate neighbourhood, the optimum partition is P^{**}. Now, if

$$Q^D(P^{**}) - Q_S(P^{**}) > Q^D(P^*) - Q_S(P^*), \tag{6.18}$$

then, since

$$Q_S^D(P^{**}, r^*) = Q_S^D(P^*, r^*)$$

we get

$$Q_S^D(P^{**}, r^* + \Delta r) > Q_S^D(P^*, r^* + \Delta r),$$

where $\Delta r \in (0, r^{**} - r^*)$. And this contradicts the assumed optimality of P^* to the right of r^*.

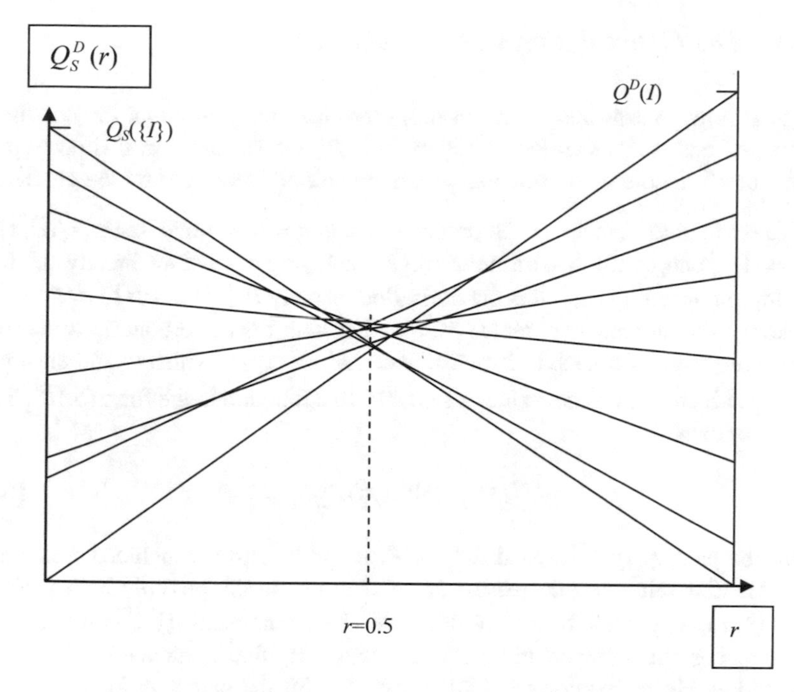

Fig. 6.3 An illustration for the outlook and formation of the function $Q_S^D(r)$

An analogous reasoning can be also carried out for the partition P^{***}, which corresponds to the maximum of $Q_S^D(r)$ for r immediately to the right of r^{**}.

If we note, additionally, that only a small proportion of the partitions $P \in E_P$ takes part in the establishment of $Q_S^D(r)$, then the schematic illustration of Fig. 6.3 becomes quite realistic.

Let us emphasise at this point that, in general, $\min_r Q_S^D(r)$, whose existence and uniqueness result from the convexity of this function, has nothing to do with the sought $\max_{P \in E_P} Q_S^D(P)$. The relation between the two functions is established through the definitions, from which (6.17) results.

It should also be remarked that Property 4 could have been derived more directly, on the basis of the properties of the function of maximum from a family of linear functions, but in view of the importance of this feature (along with Property 3) for the derivation of the algorithm and the analysis of its characteristics, the reasoning that we used here has been provided at length.

6.2.2 The General Form of the Algorithm

We shall now present the algorithm proposed, assuming only that the conditions, formulated before, concerning $Q^D(P)$ and $Q_S(P)$, are satisfied. In this description, P^* and/or P^T denote the partitions, which are selected (formed) by the algorithm.

1. We set the initial index of the procedural step t: = 0; we maximise $Q_S^D(P,r)$ for $r^t = 1$, which, conform to formulae (6.12), (6.13), means that we directly obtain the partition, resulting from this maximisation, namely $P^*(r^0) = P^*(1) = P^0 = I$.
2. Having the previously obtained P^t, together with the corresponding value of r^t, we now search for such a pair of clusters, $A_q^t, A_{q'}^t \in P^t$, which would ensure the biggest increment of the value of the objective function in the form (6.10). Thus, we compare

$$Q_S^D(P^t,r) \text{ with } Q_S^D(P_H^t(q,q'),r) \tag{6.19}$$

for the pairs q, $q' \in K_{P^t}$ and for $r < r^t$. It can be easily concluded from (6.10) that as the value of r decreases, in order to obtain the possibly high values of $Q_S^D(P,r)$, one should try to decrease the component Q^D, while possibly increasing the component Q_S, and preserving the character of P being a partition. Hence, conform to (6.19), we look for the pair q,q', for which

$$\begin{aligned} &rQ^D(P^t) + (1-r)Q_S(P^t) \\ &= rQ^D(P_H^t(q,q')) + (1-r)Q_S(P_H^t(q,q')) \end{aligned} \tag{6.20}$$

is satisfied for a possibly high value of the parameter r. Denote the value of this parameter, satisfying (6.20) for the pair q, $q' \in K_{pt}$ by $r^{t+1}(q,q')$.
Then,

$$\begin{aligned} &r^{t+1}(q,q') \\ &= \frac{Q_S(P_H^t(q,q')) - Q_S(P^t)}{Q_S(P_H^t(q,q')) - Q_S(P^t) + Q^D(P^t) - Q^D(P_H^t(q,q'))} \end{aligned} \tag{6.21}$$

and

$$r^{t+1} = \max_{q,q' \in K_{pt}} r^{t+1}(q,q') = r^{t+1}(q^*,q^{**}). \tag{6.22}$$

3. We aggregate clusters A_{q*} and $A_{q**} \in P^t$, defined through (6.21) and (6.22), and thereby we form the partition P^{t+1}.
4. We increase the step number, $t : = t + 1$.
5. If $t < n - 1$ then we return to step 2° of the procedure, otherwise we terminate the procedure.

This general procedure, which does not refer to the particular form of the objective function, but only to its general formulation, can be implemented in practical terms for the concrete bi-partial implementations, satisfying the conditions, specified in the preceding section.

Let us add, for formal completeness, that in case in formula (6.21) the denominator (and therefore also the numerator) are equal zero, this corresponds to $r^{t=1}(q,q') = 1$.

6.3 Some Comments on the Procedure

6.3.1 Divergence from Optimality

It can be easily noted what the proposed procedure lacks with respect to the potential full optimisation. Namely, the sole operation that is performed for the value of r decreasing from 1 down to the ultimate r^{n-1} is the aggregation of pairs of clusters. This means that we consider (actually: form through the algorithm) not the previously introduced function $Q_S^D(r)$, see (6.16), but another, simplified function, say $Q_S'^D(r)$, which is composed of at most $n-1$ segments of the straight line, corresponding to the consecutive P^t. There are, of course, at least two common points of $Q_S^D(r)$ and $Q_S'^D(r)$, namely

$$Q_S^D(1) = Q_S'^D(1) = Q^D(I) \text{ and } \quad Q_S^D(0) = Q_S'^D(0) = Q_S(\{I\}).$$

In order to secure an essential enhancement of the probability of achieving the actual optimum it would suffice to retain, at each iteration t of the procedure not just one r^t and the corresponding A_{q*} and $A_{q**} \in P^{t-1}$, but a set of parameter values, $\{r_v\}^t$, and corresponding cluster pairs, ranked highest. In this manner a tree would be generated, but its growth in terms of width would be limited by some bound on the quality of solutions at the given node (i.e. with the quality falling down beyond a certain limit, the branch would be discontinued).[1] All in all, it is obvious that the optimum corresponds to some node of a branch of such a tree, and the ampler the sets $\{r_v\}^t$, the higher the probability that we indeed determine this optimum.

[1]We do not treat separately here the case of multiple pairs of clusters, satisfying (6.22), since in procedural terms this appears to be of secondary significance.

6.3.2 Analogous Transformation of the Relational Form of the Objective Function

The previously formulated relational objective function variant, see (5.1–5.2), originally proposed by Marcotorchino and Michaud, can also be expressed in the form analogous to the one that we consider here, i.e. parameterised through r. Due to the transformation, we obtain the following objective function of the respective mathematical programming problem:

$$\max_{P \in E_P} \sum_{i,j \in I} ((1-r)s_{ij}y_{ij} + rd_{ij}(1-y_{ij})) \tag{6.23}$$

where we omit the constraints, brought in before, since we assume that maximisation is performed over $P \in E_P$.

As before, we change the value of the parameter r from 1 down to 0, or some other value, at which the procedure would effectively terminate (i.e. we would have formed a single, all-embracing cluster). Now, exactly as before, if we put $r^0 = 1$, then, obviously, the maximum of (6.23) is attained for $y_{ij} = 0$ $\forall i,j \in I$, meaning, identically as before, that $P^0 = I$.

Similarly as in the procedure proposed—switches from P^t to P^{t+1} would occur at the discrete values of r^{t+1}, and, after a finite number of such iterations, the procedure would end with $P^*(r^T) = P^T = \{I\}$, for sufficiently small r^T.

This realisation of the parameter-based procedure would have the obvious advantage of obtaining the actual optimum, although, as already mentioned before, larger tasks (in terms of n) might easily get prohibitively great dimensions due to the transitivity constraints.

On the other hand, however, it should be noted that for smaller tasks, manageable from the point of view of computing resources, the parameterised variant of (5.1–5.2) also presents certain advantages, when compared to the straightforward solving of (5.1–5.2). Namely, it allows for a relatively easy determination of the set of optimal solutions and those very near to the optimum, meaning the ones obtained for both $r = 1/2$ and the values of r close to 1/2. Then, it produces, together with the consecutive P^t, the corresponding values of r^t, thereby allowing for the assessment of the "ranges of validity" (i.e. $[r^t, r^{t+1}]$) of particular P^t, including the optimal and near-to-optimal ones. These ranges of validity make it possible, as well, to assess the sensitivities of the particular solutions, which is especially important for those close to and in the optimum (we shall yet return to this issue in Sect. 6.3.4).

Besides, the stepwise solution of the parameterised form often calls for comparable computational resources with the straightforward solving of the problem. This is so, because the starting solution, P^0, is trivial, and finding of the next solution is in a vast majority of cases (t's) quite simple. The variant with parameterisation, which was not known to the original authors of the formulation (5.1–5.2), i.e. Marcotorchino and Michaud, was proposed by the present author in Owsiński and Zadrożny (1986).

6.3.3 The Complexity of the Procedure

The procedure here introduced is, to a large extent, analogous to the classical hierarchical progressive merger (or agglomerative) algorithms. Moreover, for some of the concrete implementations of the procedure introduced, it is possible to obtain the formulae identical with or very similar to those developed by Lance and Williams (1966, 1967), characterising various classical algorithms of this kind.

This feature of the procedure is insofar important as it ensures that in the worst case the number of essential operations (function calculations) is $O(n^3)$. On the top of this, just like in some of the hierarchical aggregation algorithms, there is still room for simplifications for selected individual implementations of this general algorithm.

6.3.4 The Character of the Results and the Values of Parameter r

The partitions P^t, obtained from the procedure, form the dichotomous hierarchy. The hierarchy, though, is itself only a by-product of the procedure, and, in particular, the properties, concerning the objective function, do not apply to the hierarchy. A very important aspect of the results, on the other hand, is constituted by the non-increasing sequence of values $\{r^t\}$, whose elements, for $t > 0$, correspond to the consecutive operations of aggregation of pairs of clusters. Hence, the parameter r constitutes the so-called index of hierarchy, which is obtained in a natural way from the working of the procedure. At the same time, it is the parameter of the method, referred to in the here formulated properties.

The suboptimum solution that we look for is obtained, of course, for such r^{t^*} that $1/2 \in \left[r^{t^*+1}, r^{t^*}\right]$. This means that the solution might not be unique, and indeed, it is not unique when $r^{t^*+1} = r^{t*} = 1/2$, but, of course, also when either r^{t^*+1} or $r^{t*} = 1/2$.

Further, the length of the intervals $[r^{t+1},r^t]$ yields a kind of measure of sensitivity, or validity, of the partitions P^t (the bigger $r^t - r^{t+1}$, the less sensitive the solution P^t). Even though the character of the function $Q_S^D(r)$ suggests that it is natural for these interval lengths to be bigger closer to the ends of the entire interval [0,1] of r, it is still obvious that $r^t - r^{t+1} = 0$ with $r^{t+1} - r^{t+2} > 0$ suggests that we might (actually: should) simply omit the actually "virtual" solution P^t and replace it with P^{t+1}.

With respect to the values of r^t it ought also to be noted that after we start with $r^0 = 1$, the subsequent r^t shall remain equal 1 as long as (6.3) is true for some pairs of clusters, (q,q'), which can be aggregated in the consecutive iterations of the procedure.

Naturally, in view of the form of (6.21) and (6.1–6.2) it is certain that $r^t \in [0,1]$. Further, (6.5) together with (6.22) ensure that the sequence $\{r^t\}$ is non-increasing.

6.3.5 *Possibility of Weakening of the Assumptions*

The first of the possibilities of weakening the assumptions, admitted until now for the development of the general algorithmic procedure, is the abandonment of the monotonicity of the sequence of r^t. While retaining the assumption of the "opposite monotonicity" of the global components of the objective function, we can still easily notice that the general procedure can work equally well, when the sequence $\{r^t\}$ is not monotone.

Assume, namely, that in the course of the procedure we have obtained the series of values of the parameter, $r^0, \ldots, r^{t-1}, r^t$, which constitute a non-increasing, monotone subsequence of the entire sequence $\{r^t\}_0^{n-1}$. Further, assume that the subsequent value, which is obtained, i.e. r^{t+1}, is bigger than the preceding one, $r^{t+1} > r^t$. As we refer to the previously presented Figs. 6.2 and 6.3 and to their interpretation, we can easily state that starting from the current value r^{t+1} the parameterised objective function is defined by the partition P^{t+1}, and not by P^t, or, possibly, other, earlier partitions, which correspond to the values of $r \leq r^t$.

In this manner, for each consecutive value of r^t there may occur an "occlusion" (invalidation) of the previously obtained values of the parameter r and of the corresponding partitions, which obviously leads to a simplification of the form of the objective function $Q_S^D(r)$, obtained effectively through the algorithm (lesser number of the segments of the straight line, corresponding to selected partitions, contributing to the objective function). Yet, conform to the definition of $r^t(q,q')$ we are certain that its value shall not exceed 1.

The second possibility of weakening of the assumptions made until now refers to the basic requirement of the "opposite monotonicity". The procedure can, namely, work properly, when conditions (6.1–6.2) are fulfilled in the consecutive iterations of aggregation for a definite subset of pairs $(q, q') \in K_{P^t} \times K_{P^t}$. The algorithm should then be complemented, at each aggregation operation, with the stage of verification whether the currently selected pair of clusters satisfies the "opposite monotonicity" condition, and in case no such pair can be found, the algorithm ought to be terminated.

6.3.6 *Properties of the Method*

It can be easily noticed that the method, considered here as encompassing both the formulation of the bi-partial objective function and the here outlined algorithm, satisfies the conditions that one would like a clustering method to fulfil, namely:

i. the objective function, preferably allowing for the comparison of quality of partitions over the entire E_P;
ii. the effective algorithm for optimisation with respect to this objective function;
iii. the possibility of choosing various definitions of distance and similarity between the objects, between the clusters, and inside the clusters;
iv. the simplicity of operation and interpretation.

Of these four qualities, some problems arise solely—exactly—in connection with the algorithm (point ii. above), since the one here outlined leads to suboptimisation, and yet the estimation of the optimality gap (the difference with respect to the true optimum) can hardly be obtained. Still, these shortcomings are largely compensated by the advantages, offered by the procedure, commented upon before, including the simplicity and the intuitive appeal of the algorithm, as well as the possibility of easy determination of various components, involved in the algorithm (global objective function, its global components, parameter r).

Now, we shall make a kind of reference to the Sect. 5.4.4, where we considered some basic precepts serving the assessment of the quality (and/or correctness) of the partitions and the clustering methods. Namely, we shall formulate some properties that a method, which uses the parameter r, in very broad sense corresponding to the perception of distance vs. similarity ("scale"), perhaps at various levels of perception, ought to satisfy. And we shall comment on whether and how the here introduced procedure satisfies them.

Property of minimum similarity.

If $s_{ij} > 0 \ \forall \ i, j \in I$, then there exists such a value of the parameter of the method, r, i.e. $r(s^{\min})$, where $s^{\min}$ is the smallest similarity between the objects $i, j \in I$, that for $r < r(s^{\min})$ the method determines $P^*(r) = \{I\}$.

Property of minimum distance.

If $d_{ij} > 0 \ \forall \ i, j \in I$, then there exists such a value of the parameter of the method, r, i.e. $r(d^{\min})$, where $d^{\min}$ is the smallest distance between the objects $i, j \in I$, that for $r > r(d^{\min})$ the method determines $P^*(r) = I$.

Property of equal distances.

If in a given data set for every $i \in I$ there is $\min_j d_{ij} = d$, then there exists such a value of r, namely $r(d)$, that for $r < r(d)$ the method determines $P^*(r) = \{I\}$, and for $r > r(d)$ it determines $P^*(r) = I$, while for $r = r(d)$ it does not provide the basis for the choice of the partition.

Property of the ideal structure.

If the set of objects, I, can be partitioned into subsets A_q in such a way that every s_{ij} for the objects i, j belonging to the same A_q is bigger than any s_{ij} for the objects i, j belonging to different A_q's, then there exists such a value of the method parameter r, fulfilling the inequality.

$$\max_{i,j:q(i)\neq q(j)} r(s_{ij}) < r < \min_{i,j:q(i)=q(j)} r(s_{ij}) \tag{6.24}$$

where q(i) is the index of the subset (cluster) containing object i, that for this value of r the best partition of I, conform to the method, is exactly $P^*(r) = \{A_q\}$.

Let us now comment on these properties for the case of the general procedure here introduced.

Property of minimum similarity.

This property is satisfied for r^{n-1}, since $s_{ij} > 0$ implies $d_{ij} < +\infty \ \forall \ i,j \in I$, and thus also $r^{n-1} > 0$.

Property of minimum distance.

This property is satisfied at the beginning of the procedure, since following $r^0 = 1$, if $d_{ij} > 0 \ \forall \ i,j \in I$, then $r^1 < 1$.

Property of equal distances.

Fulfilment of this property depends upon the particular form of the objective function—some of the known forms, also formulated in this volume, fulfil it, although there are also such ones that do not. Thus, for instance, the form equivalent to the basic (5.1) would require a significant relaxation of this property in order to fulfil it.

Property of ideal structure.

The situation is similar as before, depending upon the concrete form of the objective function. Here, however, the form (5.1) does fulfil it.

6.3.7 The Values of the Objective Function

Another very significant advantage of the method is constituted by the fact of generating the sequences of values of Q_S^D, Q^D and Q_S for the consecutive partitions P^t. These values, obtained from the procedure, provide a valuable analytical material for the substantive assessment of the results (partitions P^t), which the procedure yields. We know of the sequences $\{Q^D\}^t$ and $\{Q_S\}^t$ that the former is non-increasing, while the latter is non-decreasing. It can be postulated—and, indeed, intuitively this might be justified—that the two components for r^t in the neighbourhood of 1/2 have similar weights, i.e. that they have similar values in the vicinity of the maximum for the original objective function. This, of course, is not so easy to ensure a priori. The best way to try to fulfil this condition is to normalise the values of distances and similarities so as to attempt ensuring the equality

$$Q^D(I) = Q_S(\{I\}) \tag{6.25}$$

which, for some definite forms of the objective function may turn out not to be very demanding.

Let us also note that the here mentioned sequence of values of $Q_S^D(P^t)$ is determined directly from the definition, but it is also being approximated by the procedural implications of the formula (6.16). The proper values of $Q_S^D(P^t)$ attain the maximum for the value of r close to 1/2, while the values of $Q_S^D(r)$ from (6.16) attain their minimum for the value of r, depending upon the concrete forms of $D(.,.)$, $S(.,.)$ etc. Thus, it can be ultimately stated that the procedure generates two kinds of essential indices of hierarchy, namely r^t and $Q_S^D(P^t)$.

6.4 Concrete Realisations of the General Algorithm

6.4.1 Introductory Remarks

In this section, we shall first present three instances of algorithmic realisations. Based on these three instances we shall comment on the characteristics of the algorithms, deduced from the general form of the bi-partial objective function, and then some other algorithmic realisations shall be shortly presented. Finally, we shall provide some principles of construction of the algorithms and an instance of analysis of selected features of the algorithms.

When presenting the concrete algorithmic realisations of the method, we shall be referring to the diverse definitions of the magnitudes, which are involved in the general description of the algorithm, given before in this chapter. This means, primarily, the definitions of $Q^D(P)$, $Q_S(P)$, $D(A_q,A_{q'})$ and $S(A_q)$, or, for the dual form of the bi-partial objective function, $Q_D(P)$, $Q^S(P)$, $D(A_q)$ and $S(A_q,A_{q'})$. On the other hand, as we reach deeper into the levels of perception—it is in principle of no importance for the procedure what definitions of distance and/or proximity between individual objects we adopt.

As concerns the particular forms of the bi-partial objective function, for which we shall formulate the algorithmic realisations, we shall also be referring to Chap. 5 , and especially to Sect. 5.3, where examples were provided of the different concrete forms of the bi-partial objective function.

Just for the sake of completeness we shall recall here the fundamental formula of the algorithmic procedure, namely the one, defining the value of $r^t(q,q')$. This is the value of the parameter of the method, corresponding to the aggregation of clusters A_q^t and $A_{q'}^t$ from among those forming the current partition P^t, generated by the procedure, i.e.

$$r^{t+1}(q,q') = \frac{Q_S(P_H^t(q,q')) - Q_S(P^t)}{Q_S(P_H^t(q,q')) - Q_S(P^t) + Q^D(P^t) - Q^D(P_H^t(q,q'))} \tag{6.21}$$

Concrete definitions of the respective functions, associated with the given form of the objective function, are being applied in the formula (6.21) to yield the way of calculating the values of $r^t(q,q')$ and hence of r^t for the corresponding algorithmic realisation.

In view of the fact that an important part of the algorithmic realisations, which are presented here further on, has the character similar to that of the hierarchical merger algorithms, based on the minimum distance rule, also those following the Lance-Williams formula, we shall repeat here, as well, this formula, which defines the distance between two clusters that have been aggregated, A_q^t and $A_{q'}^t$, on the one hand, and any other cluster, denoted here by B:

$$\begin{aligned} D(A_q^t \cup A_{q'}^t, B) \\ = a_1 D(A_q^t, B) + a_2 D(A_{q'}^t, B) + a_3 D(A_q^t, A_{q'}^t) \\ + a_4 \left| D(A_q^t, B) - D(A_{q'}^t, B) \right| \end{aligned} \tag{6.26}$$

where $a_1,\ldots,a_4$ are the coefficients, which define the actual version of the new distance between clusters.

6.4.2 The Algorithm with Additive Objective Function

We refer here to the basic, leading example that was referred to also in several other places in the present book (see, in particular, Sects. 3.2 and 5.1). Being also the leading example for the algorithmic realisation, this particular algorithm was extensively already described elsewhere, see, e.g. Owsiński (1984, 1990).

Let us recall that in this version we assume additive distances and proximities between the clusters:

$$D(A,B) = \sum_{i \in A, j \in B} d_{ij}, \text{ and } S(A,B) = \sum_{i \in A, j \in B} s_{ij} \tag{6.27}$$

and likewise for the inter-cluster measures:

$$S(A) = \frac{1}{2} \sum_{i,j \in A} s_{ij}, \text{ and } D(A) = \frac{1}{2} \sum_{i,j \in A} d_{ij}. \tag{6.28}$$

Then, the global components of the objective function are construed analogously:

$$Q^D(P) = \sum_{q=1}^{p-1} \sum_{q'=q+1}^{p} D(A_q, A_{q'}) \text{ and } Q_S(P) = \sum_{q=1}^{p} S(A_q), \tag{6.29}$$

while we remember that the sets $\{1,\ldots,p\}$ for P are denoted K_P.

For these definitions, the formula (6.21) takes on the form

$$r^{t+1}(q,q') = \frac{S(A_q, A_{q'})}{S(A_q, A_{q'}) + D(A_q, A_{q'})} \tag{6.30}$$

where $q, q' \in K_{P^t}$. It is easy to note that in view of (6.27) and the basic assumptions on distances and proximities for the same pairs of objects, introduced at the very beginning, instead of calculating (6.30), one can simply search at each iteration for the pair of clusters, q and q', featuring the smallest distance $D(A_q,A_{q'})$, for which the consecutive r^t would be determined.

This exactly is the principle of aggregation according to the minimum distance, being the basis of the progressive merger hierarchical algorithms, in particular those, for which the Lance-Williams formula (6.26) is valid. Given the definition of inter-cluster distance (6.27), the coefficients of the Lance-Williams formula for this algorithmic realisation are $a_1 = a_2 = 1$, $a_3 = a_4 = 0$. Such a setting of parameters is not envisaged in the classical setting of the Lance-Williams formula, but appropriately reflects the procedure as here presented.[2]

This algorithm requires, for each of the aggregation iterations $t > 0$, at most $n - t + 1$ comparisons of values (in the initial iteration, i.e. for $t = 0$, $1/2n(n - 1)$ comparisons). Hence, the computational burden in this algorithm can be estimated as being not bigger than $O(n^2)$.

Let us add also that, as indicated before, the objective function used in this case is equivalent to the parametric form of the one proposed by Marcotorchino and Michaud (see Sect. 6.3.2, formula (6.23)). Hence, as also indicated there, for appropriately small tasks (in terms of n), the results from the two approaches can be directly compared.

6.4.3 *The Algorithm for the Objective Function with Additive Proximities and Constant Cluster Setup Cost*

In this case, we assume the form of $Q_S(P)$ and $S(A_q)$ as implied by (6.28), but $Q^D(P)$, expressing here the "cluster setup cost", has the form, after an appropriate normalisation

$$Q^D(P) = p_P - \mathrm{card}P \tag{6.31}$$

[2]Actually, the very procedure is entirely analogous to that of "average link" hierarchical merger, as mentioned already in Sect. 5.4.3, although the parameters for "average link" are $a_1 = n_q/(n_q + n_{q'})$, $a_2 = n_{q'}/(n_q + n_{q'})$, $a_3 = a_4 = 0$, where n_q, $n_{q'}$ are the cardinalities of clusters A_q and $A_{q'}$, respectively, these coefficients expressing the fact of weighing the numbers of objects in the clusters to obtain the averages.

from which we get

$$r^{t+1}(q,q') = \frac{S(A_q, A_{q'})}{S(A_q, A_{q'}) + 1} \tag{6.32}$$

where A_q, $A_{q'}$ are the clusters, belonging to the partition P^t, established in the preceding iteration of the algorithm.

Like before, for this formulation, the search for $\max_{q,q'} r^t$, expressed through (6.32), can be replaced by the much simpler search, in this case of $\max_{q,q'} S(A_q, A_{q'})$. Given the interrelation between s_{ij} and d_{ij} this again leads to the aggregation according to minimum distance. For this algorithm the Lance-Williams formula takes identical form as for the preceding one, but it is applied solely to $S(A_q, A_{q'})$ (were it also applied to $D(A_q, A_{q'})$, the two algorithms would have to be identical).

As already noted in Sects. 4.4 and 5.3.2, this form of the objective function might arise quite naturally in the context of location problems, where proximity inside the cluster is calculated, e.g., on the basis of transport costs in relation to some benchmark, while the "cluster setup cost" is derived from the cost of constructing and maintaining a dispatching centre.

6.4.4 *The Algorithm of the Extreme Distances and Proximities*

This example will be a bit more complicated in numerical terms. First of all, it is non-additive, which, however, will not make it impossible to derive the basic elements of the procedure. The complication will be associated, on the other hand, with the ultimate form of $r^t(q,q')$, hence of the very core of the algorithm.

So, let us assume, in quite a standard manner,

$$D(A,B) = \min_{i \in A, j \in B} d_{ij} \text{ and } S(A) = \max_{i,j \in A} s_{ij} \tag{6.33}$$

with

$$Q^D(P) = \sum_{q=1}^{p-1} \sum_{q'=q+1}^{p} D(A_q, A_{q'}) \text{ and } Q_S(P) = \sum_{q=1}^{p} card\, A_q S(A_q). \tag{6.34}$$

In order to obtain the possibly most convenient form of expression for r, related to this objective function, we shall first write it out, conform to the definition, as

$$r^{t+1}(q,q') = \frac{\Delta Q_S(q,q')}{\Delta Q_S(q,q') + \Delta Q^D(q,q')} \tag{6.35}$$

which is a slightly simplified form of the initial formulation, and we shall refer to it further on (remembering, naturally, that q and q' belong to K_{P^t}).

For the here considered algorithm, the quantities, appearing in formula (6.35) are as follows:

$$\Delta Q_S(q,q') = \left(\text{card}A_q + \text{card}A_{q'}\right)S(A_q \cup A_{q'}) \\ -\text{card}A_q \cdot S\left(A_q\right) - \text{card}A_{q'} \cdot S\left(A_{q'}\right) \tag{6.36}$$

and

$$\Delta Q^D(q,q') = D\left(A_q, A_{q'}\right) \\ + \sum_{q^* \neq q,q'} \max\{D(A_q, A_{q^*}), D(A_{q'}, A_{q^*})\}. \tag{6.37}$$

Along with the increased complication of the algorithmic formulae, the computational effort increases, obviously, as well. This means that the algorithm, presented here, associated with the objective function, defined by (6.33) and (6.34), should be used only when this particular form of the objective function is really justified by the substantive or technical context of the given problem. Let us note, however, that the complexity of this algorithm is not higher than for quite some of the known clustering algorithms.

Notwithstanding this relative complexity of the involved expressions, it is possible to determine for this algorithm the Lance-Williams formula, although in this case it will be two separate formulae, one for $D(A,B)$ and another for $S(A)$.

Thus, given the formula (6.37), it can be concluded that for the inter-cluster distance $D(A_q,A_{q'})$ the values of the coefficients in the Lance-Williams formula would be: $a_1 = a_2 = 1/2$, $a_3 = 0$, and $a_4 = -1/2$, that is—exactly as for the classical "nearest neighbour" or "single linkage" hierarchical merger algorithm. Now, regarding $S(A_q \cup A_{q'})$, appearing in (6.36), which is not represented by the classical formula, we can propose an analogous expression, namely

$$S(A_q \cup (A_{q'} \cup A_{q''})) \\ = b_1 S((A_q \cup A_{q'}) + b_2 S(A_q \cup A_{q''}) \\ + b_3 S(A_{q'} \cup A_{q''}) + b_4 \left|S(A_q \cup A_{q'}) - S(A_q \cup A_{q''})\right|. \tag{6.38}$$

For such a formula, regarding intra-cluster proximity, corresponding to aggregation of two clusters (indexed q and q'), the coefficients, representing the here considered algorithm would be $b_1 = b_2 = b_4 = 1/2$, and $b_3 = 0$, like for the classical "complete link" ("farthest neighbour") algorithm.

The here outlined three algorithmic realisations, corresponding to definite forms of the bi-partial objective function, appear to be quite representative of an important class of such realisations, which are both feasible and rational from the point of view of applications. The representativeness, anyway, concerns not so much the entire formulations, as the essential component parts, especially the definitions of

$D(.,.)$, $S(.,.)$, $S(.)$, Q^D, Q_S, etc. It is therefore possible to take advantage of this fact by constructing other forms of the bi-partial objective function, using these component parts, and obtaining yet other algorithms. This is insofar important as the method proposed does not lead to a concrete prescription for the algorithm, but only to the general precepts, here provided and illustrated with definite realisations.

6.5 The Properties of the Algorithmic Realisations

6.5.1 Aggregation According to Minimum Distance

As this can be seen from the examples here provided, for various forms of the bi-partial objective function one can obtain similar—at least under some aspects—forms of the algorithmic formulae. This concerns, in a particular manner, the association with the hierarchical merger algorithms.

It must be noted, though, that the potential capacity of formulating a relation of the Lance-Williams kind in the sense of formula (6.26) or (6.38) does not mean yet that the respective algorithm is performed through the search for the mutually closest clusters (i.e. aggregation according to minimum distance).

On the other hand, however, even if an algorithm, derived within the framework of the method here introduced, can be in its entirety expressed through the Lance-Williams scheme (i.e. both the formula (6.26) and/or (6.38) and the merger of the closest clusters), the use of the method of bi-partial function gives, additionally, the possibility of calculating the values of the functions $Q^D(P^t)$, $Q_S(P^t)$, etc., and of comparing them, along with determination of the values of r^t that may suggest the suboptimal solution and characterise the properties of the generated dichotomous hierarchy. On the top of this, there exists a possibility of introducing procedures of local improvement of results (e.g. by allowing additional operations, keeping in memory of the second and third ranked closest clusters and so on).

Aggregation according to minimum distance, meaning merging of the pairs of the closest clusters, is the foundation for the vast class of hierarchical aggregation algorithms. This kind of aggregation appears also in many of the algorithms, being the realisations of the here introduced method. It is, for instance, directly implied by the formula (6.30), in connection with the respective properties of the involved functions, but this, as said before, does not mean that these algorithms are identical with the classical ("Lance-Williams") procedures of hierarchical aggregation.

This general feature of many of the here introduced algorithms is insofar important as it leads to a simplification—instead of directly maximising in the consecutive iterations the explicit expression of $r^t(q,q')$, we minimise, over $q,q' \in K_{P^t}$, the quantity $D(A_q,A_{q'}$ (or maximise $S(A_q,A_{q'})$), which allows for avoiding of the arithmetic operations, associated with repeated determination of $r^t(q,q')$. Although we do not dispose of the ready-made recipe for deriving the formula to obtain r^t that would yield the minimum distance aggregation, given the variety of the admissible forms of the objective function, it is possible, for a given kind of problem, to try out obtaining such a form of expression for r^t. Notwithstanding this statement, we shall present in the next subsection the sufficient conditions—even though not constructive ones—to obtain the minimum distance aggregation for the objective function additive with respect to clusters.

6.5.2 Objective Function Additive with Respect to Clusters

This objective function is defined by the formula (6.29), irrespective of the definitions of $D(.,.)$ and $S(.,.)$. Let us note that all of the here analysed algorithmic realisations are in this sense additive (the definitions of $S(A)$ and $Q_S(P)$, provided in (6.33) and (6.34) can, naturally, be easily appropriately transformed). The formula (6.35) presents the relation, which preserves its validity for the additive objective function, with, in general case,

$$\Delta Q_S(q,q') = S(A_q \cup A_{q'}) - S(A_q) - S(A_{q'}) \tag{6.39}$$

while

$$\Delta Q^D(q,q') = \sum_{q^* \neq q,q'} D(A_{q^*},A_q) + \sum_{q^* \neq q,q'} D(A_{q^*},A_{q'}) - \sum_{q^* \neq q,q'} D(A_{q^*},A_q \cup A_{q'}) + D(A_q,A_{q'}) \tag{6.40}$$

where, as always, we take $q, q' \in K_{P^t}$.

Now, let us note that

$$\begin{aligned} &\text{if } \Delta Q_S(q,q') \text{ is an increasing function of } S(A_q,A_{q'}),\\ &\text{then it is a non-increasing function of } D(A_q,A_{q'}), \end{aligned} \tag{6.41}$$

and so

$$\begin{aligned} &\text{if } \Delta Q^D(q,q') \text{ is an increasing function of } D(A_q,A_{q'}),\\ &\text{then, given } (6.41), (6.39) \text{ and } (6.40), \text{ as well as} (6.35),\\ &r^{t+1}(q,q') \text{ is a decreasing function of } D(A_q,A_{q'}). \end{aligned} \tag{6.42}$$

It should be mentioned that the dependences, referred to in the above conditions, namely of $\Delta Q_S(q,q')$ with respect to $S(A_q,A_{q'})$ and of $\Delta Q^D(q,q')$ with respect to $D(A_q,A_{q'})$, are meant not only as dependences upon the very values of these arguments, appearing in the respective formulae, but also as dependences upon the choice of pairs (q,q').

The sufficient conditions, formulated as (6.41) and (6.42), for the objective function to allow for the aggregation according to minimum distance, do not

Table 6.1 Selected examples of algorithms for the instances of formulations of Q^D and Q_S additive with respect to clusters

$S(A_q)$: $D(A_q,A_{q'})$:	Additive, (6.27), (6.28)	Maximum cardinality, (6.33), (6.34)	Minimum cardinality, (6.52)	Sum of means, (6.55)
Additive, (6.27)	Sect.6.4.2	ΔQ_S: (6.36) $\Delta Q^D = D(q,q')$	Sect. 6.6.5	Sect. 6.6.6
Mean distances, (6.44)	Sect. 6.6.1	ΔQ_S: (6.36) ΔQ^D: (6.46)	ΔQ_S: (6.53) ΔQ^D: (6.46)	ΔQ_S: (6.57) ΔQ^D: (6.46)
Extreme distances, (6.33)	Sect. 6.6.2	Sect. 6.4.4	ΔQ_S: (6.53) ΔQ^D: (6.47)	ΔQ_S: (6.57) ΔQ^D: (6.47)
Maximum distances, (6.48)	Sect. 6.6.3	ΔQ_S: (6.36) ΔQ^D: (6.49)	ΔQ_S: (6.53) ΔQ^D: (6.49)	ΔQ_S: (6.57) ΔQ^D: (6.49)
Partition cardinality, (6.50)	Sect. 6.6.4	ΔQ_S: (6.36) $\Delta Q^D = 1$	ΔQ_S: (6.53) $\Delta Q^D = 1$	ΔQ_S: (6.57) $\Delta Q^D = 1$

Rows and columns correspond to the definitions of $D(A_q,A_{q'}$ and $S(A_q)$, respectively, while the entries show either the subsections, where the algorithms are presented, or the formulae

represent, though, a true practical significance. This is so, because for the design of algorithms much more important is the fulfilment of conditions (6.1) and (6.2) with simultaneous preservation of the form of objective function, appropriate for the clustering task at hand. Within the thus established framework, the potential modifications of $Q^D(P)$ and/or $Q_S(P)$, meant to ensure satisfaction of (6.41) and (6.42) are practically infeasible, and so these conditions are, in fact, not constructive, while they ought to be verified in each case.

Table 6.1 shows a list of algorithms, which can be constructed with a couple of exemplary definitions of $Q^D(P)$ and $Q_S(P)$, additive with respect to clusters. These algorithms, or their elements, are briefly characterised in Sect. 6.6.

6.5.3 Additivity Along Hierarchy

It can be noticed that some of the here introduced definitions, like (6.27), (6.28) and (6.29), and also some of those appearing later on, (6.55–6.56), actually display a property that is stronger than additivity with respect to clusters. This truly strong property, which actually reduces to

$$\Delta Q^D(q,q') = D(A_q,A_{q'}) \text{ and } \Delta Q_S(q,q') = S(A_q,A_{q'}) \tag{6.43}$$

shall be referred to as additivity along hierarchy. When the bi-partial objective function features this property, then very simple expressions for r^t are obtained, such as (6.30), (6.32), or (6.51), provided further on. There may be many more of such algorithms than are described in the present chapter. And so, for instance, on the principle analogous to (6.55) it is possible to formulate $Q_S(P)$ as the sum of minima, sum of maxima etc. All of the thus obtained procedures shall lead to correct, very simple and relatively fast algorithms.

6.5.4 Distances and Proximities with Respect to Centres

The method here proposed, concerning the construction of the algorithm, can also be extended to apply for the definitions of D and S that refer to some kinds of centres or representatives of clusters, like the ones that play the major role in all the k-means like algorithms. It is obvious that the original objective functions of all the k-means-like methods are additive with respect to clusters.

In Sect. 5.2 we have shown a quite natural algorithm for the bi-partial objective function, founded on the original k-means objective function, which uses simply the basic k-means algorithm in order to obtain the suboptimal partition (the main issue being the determination of the number of clusters).

Given, however, that the objective functions of the k-means-like algorithms are additive with respect to clusters, and that their counterparts, meant to complement the bi-partial objective function, can also be formulated as such, it is possible to derive for them the complete algorithms, following the precepts, provided in this chapter. The resulting practical formulae are usually quite tedious and bring nothing new to the heart of the matter, and so we shall not present them here. It should be noted, in addition, that the complexity of the respective formulae makes the derived algorithms more demanding than those here outlined.

6.6 Algorithms for the Objective Functions Additive with Respect to Clusters

6.6.1 Algorithm of Mean Distances and Additive Proximities

In this algorithm we have $S(A)$ given as in (6.28), while $D(A,B)$ is expressed as mean distance, i.e.

$$D(A,B) = \frac{1}{cardA \cdot cardB} \sum_{i \in A, j \in B} d_{ij}. \tag{6.44}$$

It should be mentioned, though, that for this case the relations, given in formulae (6.35), (6.39) and (6.40) cannot be significantly simplified. We obtain, namely,

$$\Delta Q_S(q,q') = \sum_{i\in A_q, j\in A_{q'}} s_{ij} = S(A_q, A_{q'}) \tag{6.45}$$

just like in the basic algorithm of the additive objective function (see formulae (6.27) ff.), but also

$$\Delta Q^D(q,q') = \frac{1}{cardA_q cardA_{q'}} \sum_{q^* \neq q,q'} [cardA_{q'} \cdot D(A_{q^*}, A_q) + cardA_q \cdot D(A_{q^*}, A_{q'})]. \tag{6.46}$$

In the light of the above, this particular algorithm cannot be considered to be the one of minimum distance, but it is still quite reasonable as to its construction and the relations it is based on, while in terms of computational burden ($O(n^3)$ of function calculations) it is comparable with many other algorithms, belonging to class of hierarchical aggregation procedures.

6.6.2 Algorithm of Minimal Distances and Additive Proximities

Here, $S(A)$ is defined like in the preceding case, while $D(A,B)$—like in (6.33). Hence, the resulting $\Delta Q_S(q,q')$ is the same as in (6.45), and

$$\begin{aligned} \Delta Q^D(q,q') &= \sum_{q^* \neq q,q'} \max\left(\min_{i\in A_{q^*}, j\in A_q} d_{ij}, \min_{i\in A_{q^*}, j\in A_{q'}} d_{ij}\right) + \min_{i\in A_q, j\in A_{q'}} d_{ij} \\ &= \sum_{q^* \neq q,q'} \max(D(A_{q^*}, A_q), D(A_{q^*}, A_{q'})) + D(A_q, A_{q'}). \end{aligned} \tag{6.47}$$

6.6.3 Algorithm of Maximal Distances and Additive Proximities

In the case of this algorithm the definition of $S(A)$ is the same as before, while $D(A, B)$ is defined as follows:

$$D(A,B) = \max_{i\in A, j\in B} d_{ij} \tag{6.48}$$

that is—just like in the "complete linkage" ("farthest neighbour") algorithm from the Lance-Williams family of hierarchical aggregation algorithms. For these definitions we obtain $\Delta Q_S(q,q')$ as in (6.45), and, analogously to (6.47):

$$\Delta Q^D(q,q') = \sum_{q^* \neq q,q'} \min(D(A_{q^*},A_q), D(A_{q^*},A_{q'})) + D(A_q,A_{q'}). \tag{6.49}$$

6.6.4 *Algorithm of Partition Cardinality and Additive Proximities*

In analogy to the algorithm, described in Sect. 6.4.3 by the formulae (6.31) and (6.32), we assume

$$Q^D(P) = p_P \tag{6.50}$$

and we obtain, ultimately, also in analogy to the Sect. 6.4.3, a very simple rule for determining r^t:

$$r^{t+1}(q,q') = \frac{S(A_q,A_{q'})}{S(A_q,A_{q'})+1} \tag{6.51}$$

which means, first of all, that we deal with the minimum distance algorithm, and second—that the computational burden is relatively low ($O(n^2)$ calculations of function value).

As pointed out already before, Table 6.1 contains the summary presentation of a number of algorithms, both the ones that have been outlined here and such that can be designed with the relations here introduced. In further course of this section, therefore, we shall present explicitly only two more examples, the remaining ones being shown in Table 6.1 in terms of the way they can be constructed.

6.6.5 *Algorithm of Additive Distance and Minimum Cardinality*

Similarly to the definitions of $S(A)$ and $Q_S(P)$, provided through (6.33) and (6.34), we can introduce

$$S(A_q \cup A_{q'}) = \left(cardA_q + cardA_{q'}\right)\max\left(\min_{i,j\in A_q} s_{ij}, \min_{i,j\in A_{q'}} s_{ij}\right) \tag{6.52}$$

where q and q' indicate the clusters subject to merger. The formula for $\Delta Q_S(q,q')$, derived for these definitions, takes the form

$$\Delta Q_S(q,q') = S(A_q \cup A_{q'}) - S(A_q) - S(A_{q'})$$
$$= \left\{ \begin{array}{l} cardA_q\left(\min\limits_{i,j\in A_{q'}} s_{ij} - \min\limits_{i,j\in A_q} s_{ij}\right), if \min\limits_{i,j\in A_{q'}} s_{ij} - \min\limits_{i,j\in A_q} s_{ij} \geq 0 \\ cardA_{qq'}\left(\min\limits_{i,j\in A_q} s_{ij} - \min\limits_{i,j\in A_{q'}} s_{ij}\right), otherwise \end{array} \right\} \quad (6.53)$$

Hence, the expression for $r^t(q,q')$ has the following form:

$$r^t(q,q') = \frac{S(A_q \cup A_{q'}) - S(A_q) - S(A_{q'})}{S(A_q \cup A_{q'}) - S(A_q) - S(A_{q'}) + D(A_q, A_{q'})}. \quad (6.54)$$

6.6.6 *Algorithm of Additive Distance and Sum of Means*

In this consecutive algorithm, side by side with the additive distance, that is, the one defined by (6.27), which yields $\Delta Q^D(q,q') = D(A_q, A_{q'})$, we introduce the intra-cluster proximity (similarity), which is determined according to the following scheme:

$$S(A_q \cup A_{q'}) = S(A_q) + S(A_{q'}) + S(A_q, A_{q'}), \quad (6.55)$$

where

$$S(A_q, A_{q'}) = \frac{1}{cardA_q cardA_{q'}} \sum_{i\in A_q, j\in A_{q'}} s_{ij}, \quad (6.56)$$

and hence also

$$\Delta Q_S(q,q') = S(A_q, A_{q'}) = \frac{1}{cardA_q cardA_{q'}} \sum_{i\in A_q, j\in A_{q'}} s_{ij}. \quad (6.57)$$

For this algorithm, therefore, the formula for r^t is identical with (6.30).

6.7 Algorithms of the Objective Function Non-additive with Respect to Clusters

6.7.1 *Introduction*

It is definitely not easy to construct the functions Q^D with Q_S, or, for the dual formulation, Q_D with Q^S, that would be non-additive with respect to clusters, while, at the same time, would satisfy the conditions of the here introduced method and, first and foremost, would adequately represent the original task of cluster analysis. Let us emphasise that when we speak here of the functions that are not additive with respect to clusters, we do not mean the functions, which can be transformed into the additive ones through simple arithmetic operations, e.g. by taking logarithms.

We shall present here, therefore, three examples of the non-additive functions Q^D, which can serve to construct the algorithms for the method here introduced in conjunction with the Q_S that are additive with respect to clusters, like those defined in formulae (6.27–6.29), or other appropriate Q_S.

6.7.2 Algorithm of the Range of Inter-cluster Distance Values

The component Q^D of the bi-partial objective function has in this case the following form:

$$Q^D(P) = D^{\max}(P) - D^{\min}(P) \tag{6.58}$$

in which

$$D^{\max}(P) = \max_{q,q' \in K_P} D(A_q, A_{q'}) \tag{6.59}$$

$$D^{\min}(P) = \min_{q,q' \in K_P} D(A_q, A_{q'}). \tag{6.60}$$

It is easy to notice that in order for such a function $Q^D(P)$ to lead to an effective algorithm for solving the original clustering problem, it has, definitely, to co-appear with the function of the type of (6.27–6.29). Otherwise, it could happen that aggregation would be applied to the pair of distant clusters, provided their linkage would not exert an influence on the change of value of the entire Q.

6.7.3 Algorithm of the Sum of Distances from the Centre of Partition

In some circumstances of applying cluster analysis it may turn out useful to use the function Q^D having the form as follows:

$$Q^D(P) = \sum_{q \in K_P} D\left(A_q, x^P\right) \tag{6.61}$$

where x^P is the centre of the partition P, defined in some appropriate manner. This function is, of course, additive with respect to A_q, but not with respect to $D(A_q,A_{q'})$, as this has been assumed before, when speaking of additivity with respect to clusters. It is obvious that definition (6.61) introduces a certain additional interpretation to the original problem of cluster analysis, not present in the here adopted formulation from Sect. 6.2.1. This interpretation says that clusters are similar, or close, when they are at a short distance from the centre of partition.

Here, again, like in the preceding case, association with an appropriate definition of Q_S, like (6.27–6.29), may produce the desired effect in terms of interpretation. Notwithstanding this, though, definition (6.61) has an additional advantage that a literal counterpart can be formulated as the definition of $Q_S(P)$, being, therefore, the expression for proximity with respect to the centres:

$$Q_S(P) = \sum_{q \in K_P} S(A_q, x^q) = \sum_{q \in K_P} \sum_{i \in A_q} s(x_i, x^q) \tag{6.62}$$

where x^q are, as before, the centres or representatives of the clusters. This function is, naturally, additive with respect to clusters.

6.7.4 Algorithm of Variance of the Inter-cluster Distances

The function, which is made use of in this particular algorithm, has quite similar features to the two preceding ones. Yet, it requires an initial normalisation in order to satisfy the conditions of monotonicity, which are imposed on the components of the global objective function in the method here described. Thus, if

$$Q^D(P) = \frac{2}{(p-1)p} \sum_{q=1}^{p-1} \sum_{q'=q+1}^{p} \left(D(A_q, A_{q'}) - D^{av}(P)\right)^2 \tag{6.63}$$

where

$$D^{av}(P) = \frac{2}{(p-1)p} \sum_{q=1}^{p-1} \sum_{q'=q+1}^{p} D(A_q, A_{q'}) \tag{6.64}$$

then, for instance, for the $D(A_q,A_{q'})$, which can be determined from the Lance-Williams formula (6.26), an appropriate modification, leading to the satisfaction of the monotonicity conditions, could consist in adding 1 to the value of $D(A_{q^*}, A_q \cup A_{q'})$.

To function (6.63) we can ascribe a counterpart function $Q_S(P)$, having the interpretation, in general, of correlation, but this would lead to quite complicated relations, defining r^t.

6.8 Designing an Algorithm

As can be seen from all the previously provided examples and some more general considerations, there exists quite an ample room for manoeuvre in designing the algorithms according to the method introduced in this book. The class of functions, which satisfy the conditions (6.1) and (6.2) is definitely broad, as demonstrated by the wide choice of definitions for the components of the bi-partial objective function in both of its dual forms. On the basis of these selected instances it is possible to design some fourty algorithms of quite a diverse character. Although fulfilment of the conditions of monotonicity allows for the formulation of the algorithm according to the here adopted precepts, but it still does not guarantee the properties that would be expected from an effective algorithm. Thus, altogether, in designing an algorithm within the framework of the method, one encounters the following issues:

- fulfilment of the conditions of monotonicity with respect to hierarchy, i.e. conditions (6.1) and (6.2);
- potential fulfilment of other conditions on the objective function, facilitating the functioning of the algorithm (additivity with respect to clusters, additivity along hierarchy, aggregation according to minimum distance, applicability of the Lance-Williams formula);
- in the framework of the above—fulfilment of conditions (6.3) and (6.5), which guarantee the monotonicity of the sequence of values of r^t;
- analysis of the scales of values of the global components of the objective function and of r^t;
- evaluation of results.

Two first items above have already been commented upon, the third one is largely contained in the second, and hence we shall shortly consider only the last two points.

The analysis of the scales of values for the components of the objective function and the parameter r^t can be carried out in the context of the already taken decision as to the definition of $d(x_i,x_j)$, and before the choice of the concrete forms of $s(d)$, D, S, etc.

Let us illustrate this on the simplest, it seems, example, namely that of the definitions (6.27–6.29). First we note that in the definition of $Q^D(P)$ there appear the values of distances d_{ij}, whose total number is:

$$N_d\left(Q^D(P)\right) = \sum_{q=1}^{p-1} \sum_{q'=q+1}^{p} cardA_q cardA_{q'} \tag{6.65}$$

and in the definition of $Q_S(P)$ there appear

$$N_s(Q_S(P)) = \frac{1}{2}\sum_{q=1}^{p} cardA_q(cardA_q - 1) \tag{6.66}$$

Table 6.2 Illustration for the approximate fulfilment of condition (6.69)

n	4	5	6	7	8	9	10	11	12	13	14	15
$p^{\max}$	2	2	2	2	3	3	3	4	4	4	4	5
$n_1^{\min}$	3	4	5	5	6	7	8	8	9	10	10	11
$n_1^{\max}$	3	4	5	6	6	7	8	8	9	10	11	11

For consecutive values of n from 4 to 15 the maximum partition cardinality ($p^{\max}$), and the bounds on the cardinality of the dominating cluster ($n_1^{\min}$ and $n_1^{\max}$) are provided, for which condition (6.69) is fulfilled with sign "$\geq$", as harder to fulfil than vice versa

values of proximities s_{ij}. It is easy to note that the condition

$$N_d\big(Q^D(P)\big) = N_s(Q_S(P)), \tag{6.67}$$

which might be associated with the requirement of a possibly equal contribution of the two components to the global objective function, is equivalent to

$$\sum_{q \in K_P} cardA_q \left(\sum_{q' \in K_P, q' \neq q} cardA_{q'} - (cardA_q - 1) \right) = 0 \tag{6.68}$$

and hence also to

$$\sum_{q \in K_P} (cardA_q)^2 = \frac{n^2 + n}{2}. \tag{6.69}$$

The last condition is approximately satisfied when we deal with a partition, composed of one large, dominating cluster and a certain number of the smaller ones. This is illustrated with the examples, provided in Tables 6.2 and 6.3.

Table 6.2 makes it apparent how for small n the constraint, imposed by condition (6.69), narrows down to almost zero the possible choice of the structure of partition. Then, in Table 6.3 it is shown how for a slightly bigger n, namely for $n = 100$, the limits of cardinality of the dominating cluster change with increasing cardinality of partition, p.

The conclusion seems to be obvious: the concrete algorithm considered, based on (6.27–6.29), is from this particular point of view strongly biased. Let us, however, formulate against this background a couple of more general remarks, concerning algorithm design:

- a similar analysis ought to be performed for any other algorithm, also with respect to the values of r^t, and, in particular, with respect to the situation, arising for r^t around 1/2 (let us note that for the algorithm, analysed in this section, r^t attains the values close to 1/2 for the roughly similar structure of partition as that determined on the basis of the condition of equal weights of Q^D and Q_S, this

Table 6.3 The bounds on the cardinality of the dominating cluster, $n_1^{\min}(p)$ and $n_1^{\max}(p)$, for increasing values of p in the case of $n = 100$, ensuring fulfilment of (6.69) with the sign "$\geq$"

p	2	3	4	5	6	7	8	9	10	11	12	13	14	15	16	17	18	19	20	21	22	23	24	25	26	27	28	29	30
$n_1^{\min}$	55	59	61	62	63	64	65	65	67	67	68	68	69	69	69	70	70	70	70	71	71	71	71	71	71	71	71	71	71
$n_1^{\max}$	99	98	97	96	95	94	93	92	91	90	89	88	87	86	85	84	83	82	81	80	79	78	77	76	75	74	73	72	71

being an important, advantageous feature of this algorithm, though somewhat spoiled by the bias, described before);

- when analysing the conditions for possibly equal weights of Q^D and Q_S, the dependence upon the transformation $s(d)$ or $d(s)$ ought also to be considered;
- finally, the consistency of the two kinds of conditions (equal weights of global components and properties for the r^t close to 1/2) should be analysed, as well.

As shown throughout the book, the approach here considered allows for a wide margin of choices at various "levels of perception", starting with the very description of the objects, through definition of distances (or proximities), then the transformation between the two, the measures at the level of clusters, and finally—at the level of the entire partition. These choices ought to be driven by both the substantive and technical motivations to form a coherent wholesome approach.

Just in order to illustrate the diversity of the aspects, which ought to be considered, when designing in detail the algorithm, we shall consider here one of those, already slightly touched upon in Sect. 3.2. Namely, the choice of the transformation $s(d)$ depends, in particular, on the distribution of distances among the analysed objects.[3] Consider, for instance, four simplistic cases, shown in Figs. 6.4, 6.5, 6.6 and 6.7. The corresponding distributions of the numbers of distances are provided in Figs. 6.8, 6.9, 6.10 and 6.11.

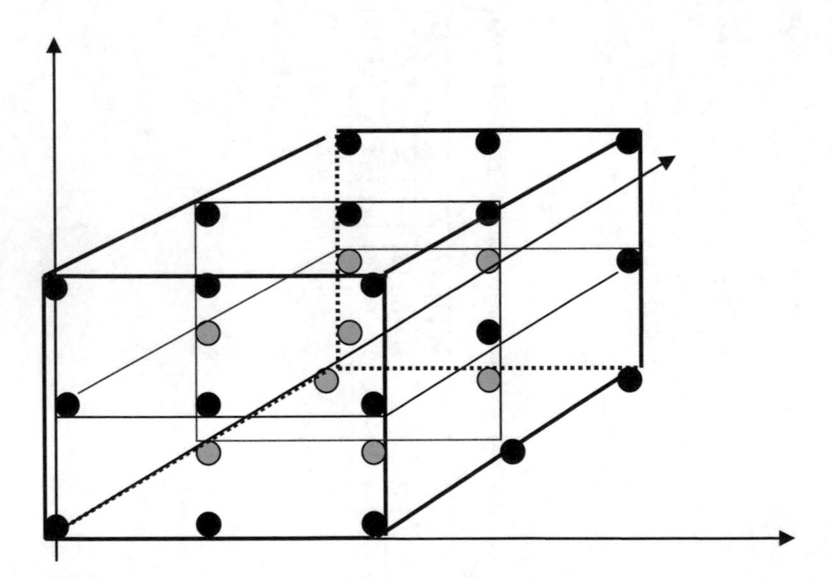

Fig. 6.4 A simplistic case of a cube, with 27 objects, positioned equidistantly on the surfaces and in the centre of the cube. The distances between the neighbouring objects are equal 1

[3]It should be kept in mind that in general, very often little attention is paid to the different „perception levels", especially the roles, played by the variable scales, definitions of distance, etc., in the clustering exercises.

Fig. 6.5 Ten objects, situated on a single line, with, again distances between neighbours equal 1

Fig. 6.6 Twelve objects, situated on a single line, with distances between the closest neighbours equal 1

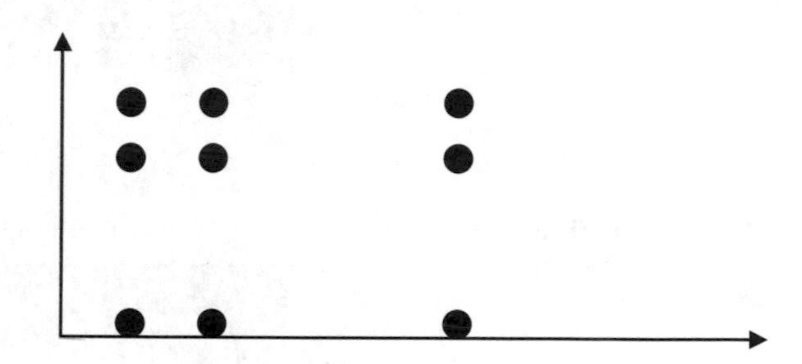

Fig. 6.7 Another simple example, again with distance between neighbours equal 1

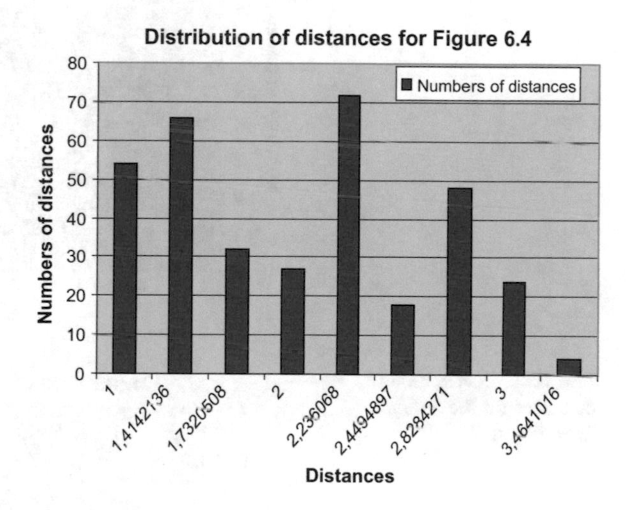

Fig. 6.8 Distribution of distances for the example from Fig. 6.4

It can be easily seen from the latter figures that (**a**) the distribution itself is not very much telling of the "shape of partition", unless one deals with very specific, "extreme" cases; but still, (**b**) it may be suggestive as to the choice of transformation $s(d)$, meaning, in particular, whether one opts for the simple transformation $s = 1 - d$ (after unitarisation), or for the one that preserves the average value of the distance, or yet another one (see, e.g., the formulae (3.8) and (3.9)).

Of special interest may be the comparison between Figs. 6.8, 6.10 and 6.11. While it is true that there exist methods (e.g. the "silhouette" type of techniques), which use the distribution of distances to construct partitions, this example shows

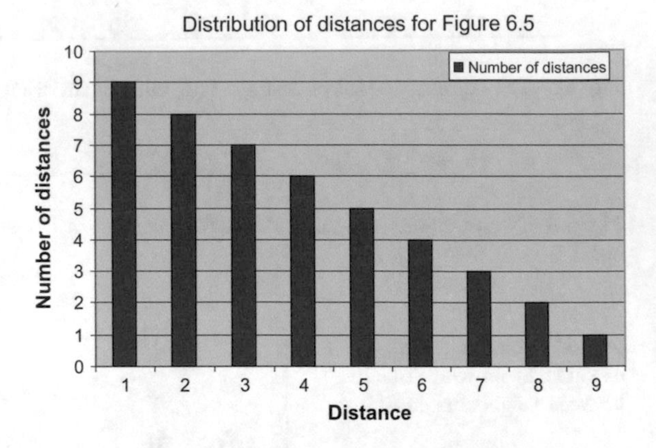

Fig. 6.9 Distribution of distances for the example from Fig. 6.5

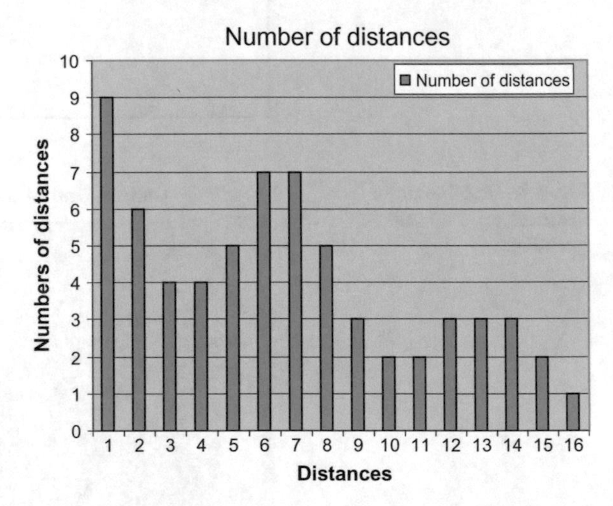

Fig. 6.10 Distribution of distances for the example from Fig. 6.6

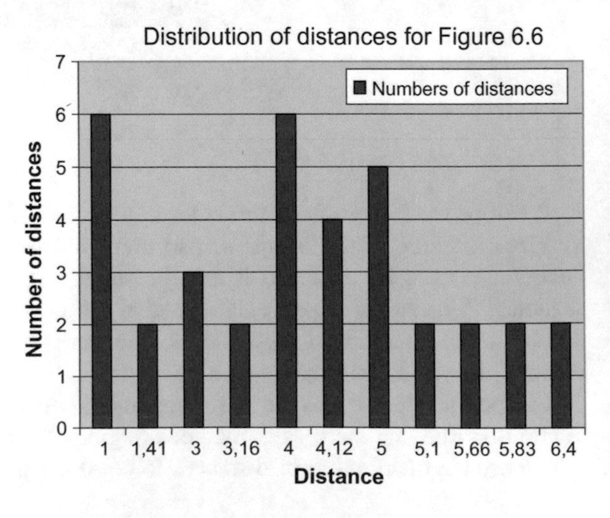

Fig. 6.11 Distribution of distances for the example from Fig. 6.7

very strongly that there must be a cost to it. On the other hand, all three examples suggest, definitely, that prior to the detailed design of the algorithm, involving the setting of all the parameters, side by side with the substantive considerations, related to the nature of the objects and the problem, the very general characteristics of the data set, including, for instance, the distance distribution, ought to be taken into account. It is generally much easier to visualise this distribution than many other aspects of such data sets.

References

Lance, G. N., & Williams, W. T. (1966). A generalized sorting strategy for computer classifications. *Nature, 212,* 218.

Lance, G. N., & Williams, W. T. (1967). A general theory of classification sorting strategies. 1. hierarchical systems. *Computer Journal, 9,* 373–380.

Owsiński, J. W. (1984). On a quasi-objective global clustering method. In: E. Diday et al. (Eds.), *Data analysis and informatics* 3 (pp. 293–305). North Holland, Amsterdam.

Owsiński, J. W. (1990). On a new naturally indexed quick clustering method with a global objective function. *Applied Stochastic Models and Data Analysis, 6,* 157–171.

Owsiński, J. W., & Sł, Zadrożny. (1986). Structuring a regional problem: Aggregation and clustering in orderings. *Applied Stochastic Models and Data Analysis,* 2(1–2), 83–95.

Chapter 7
Application to Preference Aggregation

7.1 Introductory Remarks and the Setting

This part of material in the book, an extension of the shorthand description in Sect. 4.9, is presented here as a separate chapter for two reasons: first, the problem is indeed different from the other ones, treated as examples of general character in Chap. 4 (the two opposing aspects are much less obvious), and second, it requires a justification of the setting and introduction of several specific notions and notations.

Thus, we deal, again, with n objects, which we would like to order (e.g. "from the best to the worst"), if possible—so as to form a ranking. This is a standard problem in a vast area, abundant with various models, differing as to what kinds of data are available, what assumptions are made, and, finally, what methods are used to obtain the final ranking or to state that we cannot obtain such a ranking under the given circumstances.

In this particular case we assume that each object i, $i \in I = \{1,\ldots,n\}$, is characterised by m aspects (criteria, assessments, expert opinions etc.), or, more precisely, that it **can** be characterised by **up to** m aspects. On the basis of these we are able to produce, for each pair of objects, (i, j), three numbers: σ_{ij}, corresponding to the preference of i over j (or precedence of i before j), δ_{ij}, corresponding to the preference of j over i (or precedence), and m-σ_{ij}-δ_{ij}, reflecting the measure of lack of indication as to the preference (precedence) of i and j.

A very simple instance of such a situation is provided by the "voting" of m experts, numbered $k = 1,\ldots,m$, who are asked to express their opinion on the pairs (i, j) in terms of preference of i over j ($\sigma_{ijk} = +1$) or vice versa ($\delta_{ijk} = +1$), or abstention ($\sigma_{ijk} = \delta_{ijk} = 0$). By summing over k we obtain the three numbers, mentioned before. Actually, the experts might, quite plausibly, be separately asked about their opinions as to the values of σ_{ijk} and δ_{ijk}, which would create yet an additional "degree of freedom", even if we limit somehow the possible specifications of σ_{ijk} and δ_{ijk} (a typical case being the limit on the sum of either or both of these).

J. W. Owsiński, *Data Analysis in Bi-partial Perspective: Clustering and Beyond*, Studies in Computational Intelligence 818,
https://doi.org/10.1007/978-3-030-13389-4_7

Another instance is provided by the assignment of +1 to σ_{ij} or to δ_{ij}, depending upon the comparison of the values of x_{ik} and x_{jk}, i.e. the scores of the objects i and j on the variable k, with value 0 corresponding to draws or to lack of at least one of these values, x_{ik} and x_{jk}.

Based on such data, we would like to obtain the complete ordering of the objects from I. An intuitively appealing procedure is to just sum up the respective numbers:

$$\sigma_i = \sum_{j=1}^{n} \sigma_{ij} \quad \text{or even} \quad \sigma_i = \sum_{j=1}^{n} (\sigma_{ij} - \delta_{ij}) \tag{7.1}$$

and perform ordering on the basis of decreasing values of σ_i. Yet, there is a twofold problem with such an approach, namely:

(1) we have made no assumptions on the relations between the numbers σ_{ijk} and δ_{ijk} for various i and j, and so there might be, for instance, $\sigma_{ijk} = \sigma_{jj'k} = {}_{j'ik} = +1$, meaning "$i$ better than j, j better than j', j' better than i", an apparently unacceptable situation, which may, however, arise from expert assessments; the same—or similar—may also apply to the values of σ_{ij} or to δ_{ij};
(2) the values m-σ_{ij}-δ_{ij} may significantly differ from 0, and also may be significantly different for various pairs (i, j).

Both these features of the data, obtained in the manner here described, make the simple approach proposed, invalid.

7.2 The Linear Programming Formulation

7.2.1 The Linear Program

The problem thus posed may be represented, and indeed, solved, through the formulation entirely analogous to that, which was shown in Sect. 5.1.2, and which followed Marcotorchino and Michaud (1979, 1982). Actually, these two authors also proposed this formulation for the preference aggregation problem, considered here. A treatment of this problem in the bi-partial framework is provided, as well, in Owsiński (2011, 2012).

Thus, the following mathematical programming formulation is proposed:

$$\max_V \left(\sum\nolimits_{i<j} s_{ij} v_{ij} + \sum\nolimits_{i<j} d_{ij} (1 - v_{ij}) \right) \tag{7.2}$$

under the constraints

$$v_{ij} \in \{0, 1\} \quad \forall i, j \in I, \tag{7.3}$$

$$v_{ij} = 1 - v_{ji} \quad \forall i,j \in I, \tag{7.4}$$

$$v_{ij} + v_{jj'} - v_{j'i} \leq 1 \quad \forall i,j,j' \in I, \tag{7.5}$$

where, now, $V = \{v_{ij}\}_{ij}$ are the decision variables, i.e. the ones defining the sought ordering, with $v_{ij} = 1$ corresponding to the fact that in the output ("aggregate") ordering, resulting from solving the program (7.2)–(7.5), object i precedes ("is better than") object j, while $v_{ij} = 0$ (and so, according to (7.4), $v_{ji} = 1$) corresponds to the fact that it is j that precedes i in the output ordering.

The constraints (7.3)–(7.5) ensure that the solution is, in fact, an ordering, just like the constraints (5.2) ensured that the solution to (5.1)–(5.2) was a partition.

It now becomes obvious that this program leads to a solution, being an ordering, even if there exist in the data the problems, mentioned before (i.e. local lack of transitivity and/or missing data).[1]

The question therefore arises, in a natural manner, what is the "upper bound" on these irregularities, for which one still obtains a "reasonable" solution. Since, however, there may be quite various kinds of irregularities, and also quite different assumptions as to what is a "reasonable solution", the problem definitely exceeds the frames of the present book. Let us only add that just like in the case of the problem (5.1)–(5.2), the one here formulated can be solved with the constraint (7.3) in the form of $v_{ij} \in [0,1]$, as the proof exists that the solutions to such a continuous problem are, in fact, always of binary character, i.e. $v_{ij} \in \{0,1\}$.

Another issue, which was already discussed in the clustering context in Chap. 3 [see, for instance, formula (3.10)] is that of considering simultaneously σ_{ij} and δ_{ij}. We have already partly explained the pragmatic source of this fact, but let us add at this point that by accounting for these two numbers we also account for m-($\sigma_{ij} + \delta_{ij}$), which would otherwise disappear entirely, pushing us away from the reality even farther (and this also constitutes an important difference with respect to the case of clustering).

7.2.2 The Parameterisation

Just like in the case of clustering, we can parameterise the objective function from (7.2) and thereby the entire linear program, by introducing the analogous parameter r, $r \in [0,1]$, so as to form the following objective function:

---

[1]Hence, for instance, the LP formulation presented can be used to assess the "objective" outcomes from the sporting cup-like competitions, in which not all competitors play against all the others (like in the soocer Champions' League, for instance). It can be shown that for similar cases the "reasonable solution" is usually achieved, see also the example, treated at some length at the end of the present chapter.

$$C(V,r) = (1-r)\sum\nolimits_{i<j}\sigma_{ij}v_{ij} + r\sum\nolimits_{i<j}\delta_{ij}(1-v_{ij}). \tag{7.6}$$

Now, if we assume that we solve the problem with the value of the parameter changing from one end of the interval [0,1] to the other, like in the clustering case, obtaining on the way the (finite number of distinct) solutions $V^*(r)$, we might well start from, say, $r^0 = 0$, meaning that we actually deal with

$$C(V,0) = \sum\nolimits_{i<j}\sigma_{ij}v_{ij}. \tag{7.7}$$

It is obvious that maximisation of $C(V,0)$ must lead to the solution $V^*(0)$, equivalent to the original ordering of object indices, that is, $V^*(0) = I$. This solution is, indeed, unique, provided only that all of the σ_{ij} are non-zero.

In complete analogy, for the $r^T = 1$, where T denotes the total number of distinct solutions obtained through parameterisation, i.e. for the maximisation of

$$C(V,1) = \sum\nolimits_{i<j}\delta_{ij}(1-v_{ij}), \tag{7.8}$$

under a similar condition for δ_{ij}, we obtain, as the solution $V^*(1)$, the reverse ordering to I that we denote I^{-1}.

As we consider the transformations from I to I^{-1} for the consecutive r^t, we can postulate that, similarly as for the clustering case, the shifts in the ordering $V(r)$ occur at some characteristic combinations of the values of σ_{ij} and δ_{ij} (say, analogous to the "minimum distance" for clustering). At least at the very beginning of the procedure, i.e. for the maximum value of the objective function, $C^*(r)$,

$$C^*(0) = \sum\nolimits_{i<j}\sigma_{ij} \tag{7.9}$$

we deal with a similar comparison of values of the objective function as that for clustering, i.e. we look for the minimum value of $r \geq 0$ such that

$$(1-r)\sum\nolimits_{i<j}\sigma_{ij} = (1-r)\sum\nolimits_{i<j}\sigma_{ij}v_{ij} + r\sum\nolimits_{i<j}\delta_{ij}(1-v_{ij}), \tag{7.10}$$

where the values of v_{ij} on the right hand side are no longer simply = 1. This leads to a very facile expression for $r^1(V)$, V being in this case the ordering immediately following the initial order of objects, namely

$$r^1(V) = \frac{\sum_{i<j}\sigma_{ij}(1-v_{ij})}{\sum_{i<j}(\sigma_{ij}+\delta_{ij})(1-v_{ij})}. \tag{7.11}$$

We look for the V that would minimise (7.11). Hence, it is easy to see that we, in fact, look for the possibly biggest difference $\delta_{ij}-\sigma_{ij}$ for all pairs (i, j), ordered (still) in the original sequence,[2] and for the respective pair found we exchange the places of the involved objects in the ordering (it is safer to speak of the differences than of ratios, since the lower values can be equal 0).

Naturally, the subsequent transformations of $V^*(r)$, leading to the true optimum ordering for the given value of r, can be much more complex, but, in any case, we obtain a sequence of optimum $V^*(r^t)$, and, supposedly, the one that we are after, i.e. obtained for the value of $r = ½$. Yet, as this was already mentioned for the clustering case, the sheer number of constraints, especially (7.5), being of the order $O(n^3)$, makes the practical use of the methodology for only slightly bigger instances very cumbersome.

7.3 The Alternative Algorithmic Approach

The form of the objective function, provided in (7.6), can already be treated as "bi-partial", since it contains the separate quantifications of the i before j and vice versa ordering of any two objects. We shall not introduce any new form, as the presence of the decision variables v_{ij} can be considered to simply secure the character of the solution, i.e. ordering, which will be preserved in the alternative algorithmic approach, here outlined, just by construction.

Thus, we adopt a similar perspective on the way to finding the ordering V that fits the best the available data on σ_{ij} and δ_{ij} as we did in the preceding section. Namely, we assume that the same parameterisation is introduced of the objective function, and we look for the parameter values, r^t, for which definite types of shifts in the preceding ordering, V^{t-1}, take place. We assume, like before, that $V^0 = V(r^0 = 0) = I$, given that quite mild conditions on σ_{ij} and δ_{ij}, usually holding, indeed, are fulfilled.

The respective algorithmic scheme that starts from this point and uses the approach that is similar to that illustrated in the preceding section, looks as follows:

1. set step number, $t = 0$, $r^t = 0$, and then, according to the previous analysis, the optimum ordering, $V(r^t = 0) = V^{t=0} = V^0 = \{1,\ldots,n\}$, that is—it is identical with the initial numbering of objects;
2. for all pairs of objects ordered, indexed now k and l, $k < l$, calculate two values:

$$r^t_{kl} = \frac{\sum_{i \in O^t_{l-}(k,l-1)} \sigma_{li} - \sum_{i \in O^t_{l+}(k,l-1)} \sigma_{li}}{\sum_{i \in O^t_{l-}(k,l-1)} (\sigma_{li} + \sigma_{il}) - \sum_{i \in O^t_{l+}(k,l-1)} (\sigma_{li} + \sigma_{il})}, \tag{7.12}$$

[2] In further course of the procedure—ordered according to the current V^t, which we seek to replace by the V^{t+1}.

and

$$r_{lk}^{t} = \frac{\sum_{i \in O_{k+}^{t}(k+1,l)} \sigma_{ik} - \sum_{i \in O_{k-}^{t}(k+1,l)} \sigma_{ik}}{\sum_{i \in O_{k+}^{t}(k+1,l)} (\sigma_{ik} + \sigma_{ki}) - \sum_{i \in O_{k-}^{t}(k+1,l)} (\sigma_{ik} + \sigma_{ki})} \quad (7.13)$$

where values in (7.12) and (7.13) were obtained for the equivalence conditions in the objective function (7.6) for the ordering V^t when l would be put just ahead of k in the next ordering, V^{t+1} (value of r_{kl}^t), and when k would be put just behind l in the next ordering, V^{t+1} (value of r_{lk}^t).

Notation $V_{k+}^t(k+1,l)$ refers to the set of objects in the ordering V^t having initial indices higher than k ("k+" in the subscript), which are contained in this ordering between objects k + 1 and l. Other notations are analogous.

3. find $\min_{kl}\left[r_{kl}^t, r_{lk}^t\right] = r_{k*l*}^t$ and perform the corresponding switching operation, forming thus the next ordering, V^{t+1}, t: = t+1;
4. if V^t is not the reverse of V^0 or $r_{k*l*}^{t-1} = 1$, then go to step 2, otherwise stop.

It can be easily noticed that this scheme is more complex than the one for clustering, even though also here the class of operations allowed is quite narrow (notwithstanding the fact that by such individual switches one can arrive at the hypothetical optimum solution.[3]) Let us summarise the strong and weak sides of the scheme for ordering:

- like in the case of clustering, an explicit objective function is being suboptimised, and the solution sought is obtained for the lowest r exceeding ½;
- the scheme and the procedure are simple and largely intuitive, while addressing a well-justified formulation of the problem;
- the successive "solutions" V^t are characterised by the corresponding values of r^t, whose intervals [r^t, r^{t+1}] constitute an index of validity;
- the scores, characterising the pairs (i, j) need not be exhaustive (i.e. the number of times one is "better" than another and vice versa need not add up to the assumed total of m), as indicated before, or may even be inexistent for some pairs—the place of the respective objects shall be determined by the scores with respect to other objects (in view of the reversals of indices we used in the formulae (7.12) and (7.13) only notation σ);
- further, the data used can be intransitive and inconsistent;

but:

- in general, the values of r^t obtained in the procedure may not form a non-decreasing sequence, and in case of "irregular" data, like the ones with gaps, inconsistencies etc. this becomes even more serious;

[3]Just like in the case of clustering, the true optimum could be achieved by the appropriate sequence of cluster mergers, though, in general, not necessary the one produced by the suboptimisation algorithm.

- there may even occur cycling phenomena, which have to be avoided by artificial mechanisms.

This scheme is better, perhaps, interpreted by referring directly to the objective function (7.6). In particular, it is easy to note that when either σ_{ij} or δ_{ij} is equal to zero, while the other coefficient is not, then the suboptimum shift can occur without changing the value of r from 0, respectively from 1. This corresponds to an "obvious" precedence of objects, as resulting from the direct comparison (direct in the sense that it involves only the two objects, but might result, for instance, from ranking). Yet, in further course of the procedure this shift might yet get "corrected" by consideration of relations with other objects.

7.4 An Illustration

The example that we shall present and consider here is meant to highlight as much the problem itself, as the potential various ways of interpreting it, but also, to a definite extent, the general sense of the procedure proposed in this chapter.

We shall, namely, refer to the footnote from Sect. 7.2.1, suggesting soccer cup competitions as one of the cases to be treated by the approach. Thus, we shall analyse the case of a relatively small event, in which only six teams participate, called here A, B, C, D, E and F. These six teams are split into two groups, {A, B, C} and {D, E, F}. All teams play against each other in the groups (three matches in each group), and then the teams are ordered inside the groups, as nos. 1, 2 and 3. Finally, teams ordered first in the two groups play against each other, teams ordered second against each other, and teams ordered third against each other (meaning, again, three matches). The winner of the first match takes the final winning position, the winner of the second match takes the final third position, and the winner of the third match takes the final fifth position. The losers in these three matches take, respectively, the second, fourth and sixth positions in the entire tournament. Thus, altogether nine matches are played and the ordering of the six teams, according to the here described "rules of the game", is established.

Table 7.1 shows an instance of possible outcomes of the nine matches played.

Table 7.1 An instance of the set of match scores for a small soccer cup

Teams	A	B	C	D	E	F
A	–	1:1	3:0	3:1		
B		–	2:2		1:2	
C			–			0:1
D				–	3:1	1:2
E					–	2:0
F						–

Following the "rules of the game" presented, we get for the matches of the first group the following ordering:

Teams	Match points[a]	Goal scores
A	4:1	4:1
B	3:3	3:3
C	1:4	2:5

and for the second group:

D	3:3	4:3
E	3:3	3:3
F	3:3	2:3

[a]Match points: 3 for winning, 1 for drawing, and 0 for losing

The above means that while in the first group the match points were decisive, in the second group it was the goal score that had to be considered to settle the ordering.

Then, given the results of the group matches, A played with D (group winners), winning 3:1, B with E (second teams in the groups), losing 1:2, and C with F (group losers), losing, as well, 0:1. Thereby, the final ordering of the cup was established, namely: A, D, E, B, F, C.[4]

It is all too easy to imagine that a different system were adopted to establish the final ordering. Thus, an alternative ordering could be based on match points for all the matches played by the teams, or the goal scores. A simple comparison of the resulting orderings is provided in Table 7.2.

While the orderings according to match points and goal scores are the same, they differ from the "official" results of the cup, and that in quite a significant manner (only first and last teams remain at their places). It should be emphasised that the comparison as in Table 7.2 can be made realtively easily, because all the teams played the same number of matches, which is by no means the case for all kinds of similar competition events (the eliminated teams or players playing less matches than the winners).

If matches were played between all the teams, this would make the total of 15 matches, while here altogether nine matches were played. It is quite intuitive to reason that for some kind of ordering to be made two conditions ought to hold: (a) all teams (players) have to play at least once; and (b) at least $n-1$ matches have to be played. This, of course, does not secure the "fair" treatment we mentioned in the preceding paragraph, namely that all the participants play the same number of times.

[4]The difference with the tournament, in which all teams play with each other, is, first, that instead of ½ $n(n-1)$ matches, meaning 15, only 9 are played here, but, even more importantly, the organisation of the tournament leaves less space for "strategic playing", since each match brings much more serious consequences in terms of final outcome.

Table 7.2 Comparison of three orderings: resulting from the cup "rules of the game", match points and goal scores

Teams	A	B	C	D	E	F
Cup results (ranks)	**1**	**4**	**6**	**2**	**3**	**5**
Total match scores	7:1	2:5	1:7	3:6	6:3	6:3
Ranks for match scores	**1**	**5**	**6**	**4**	**2**	**3**
Total goal scores	7:2	4:5	2:6	5:6	5:4	3:3
Ranks for goal scores	**1**	**5**	**6**	**4**	**2**	**3**

Table 7.3 Match results shown for the ordering according to the "rules of the game"

Teams	A	D	E	B	F	C
A	–	3:1		1:1		3:0
D		–	3:1		**1:2**	
E			–	2:1	2:0	
B				–		2:2
F					–	1:0
C						–

In order to just illustrate the first step of the procedure as proposed here, let us permute Table 7.1 so as to show the match results for the ordering, obtained from the cup rules (Table 7.3).

If we use directly the data from Table 7.3, the obvious first exchange of places takes place between D and F, the result of their match (in bold) being the sole score reversal with respect to the initial ordering here assumed (being an alternative "solution"). This, of course, does not mean that the next steps would not change it yet, other reversal appearing in the subsequent tables.

We shall not go deeper into the procedure, for it will get much less comprehensible for the Reader. Suffice to say that by using it we may not only obtain the results for awkward situations like this one, but also for the "unfair" ones, and that along with the corresponding values of r^t, pointing out the validity of the consecutive orderings V^t.

7.5 Some Conclusions

Just as this was mentioned at the beginning of the present chapter, we deal here with a class of problems contained in a very vast domain, in which various versions and formulations of respective problems exist, along with the corresponding models of the situation treated, and the associated assumptions, as well as proposed solution methods. Even if not addressing a number of the otherwise analysed questions, the approach proposed constitutes a relatively simple response to several issues that appear in the practical cases of the class of situations considered, such as:

inconsistency in the data on pairwise precedences, lack of individual data items, or impossibility of "neutral" normalisation of pairwise precedences. On the top of this, we dispose of the objective function and the parameter r, which allow for an assessment of the (suboptimal) solutions obtaned.

It should be stressed, though, that the range of situations, which can be treated in an effective and reasonable manner with the approach proposed is, ultimately, not so narrow. Let us only mention, for the sake of illustration, that the method is well fitted to the case, when the input data have the form of m rankings, V_k, $k = 1,\ldots,m$, with, again, the possibility that these rankings are "incomplete" (only some objects are explicitly ranked) and that there are ties. The rankings are then turned into the values of pairwise comparisons, σ_{ij}^k, with, in this particular case, $\sigma_{ij}^k = 1$ when object i precedes object j, and $\sigma_{ij}^k = 0$ in the opposite case. If we were dealing with the complete linear orders—full rankings without ties—then there would be $\sigma_{ij}^k = 1 - \sigma_{ij}^k$ for all i, j, k.

We are looking for the ordering V^*, which constitutes the "resultant" of the rankings V^k. It is quite natural to ask for V^* that would satisfy the following condition:

$$V^* = \arg\min_V \sum\nolimits_k d\left(V^k, V\right). \tag{7.14}$$

This simply means that we look for some sort of "barycentre" in the set of various rankings. The above, when expressed through the values of σ_{ij}^k, takes the form of

$$V^* = \left\{v_{ij}^*\right\} = \arg\min_{\{v_{ij}^*\}} \sum_{k=1}^{m} \sum_{i,j \in I, i<j} |\sigma_{ij}^k - v_{ij}^*|. \tag{7.15}$$

This kind of solution construct is sometimes referred to as Kemeny's median, mainly following the seminal work of Kemeny and Snell (1960) (although Bury and Wagner 2002 is the work also highly worth consulting in this context).

It can, however, be easily noticed that the problem in the form of (7.15), with the requirement on $\left\{v_{ij}^*\right\}$ to constitute a linear order (see the constraints (7.3)–(7.5) that ensure it), can quite well be solved for the input data, which do not satisfy the similarly strict conditions. In addition to this, the formulation (7.15) of the objective function appears as equivalent to another one, namely

$$V^* = \left\{v_{ij}^*\right\} = \arg\max_{\{v_{ij}^*\}} \sum_{k=1}^{m} \sum_{i,j \in I} \sigma_{ij}^k v_{ij}^*. \tag{7.16}$$

This, in turn, leads us to the possibility of summing over k and splitting the "for" and "against", regarding the precedence in each of the pairs of objects, to ultimately arrive exactly at the bi-partial formulation here proposed, that is—(7.2)–(7.5).

We have mentioned already several times over that the here considered case of aggregation of precedences (preferences, rankings, orderings,…) seems to be a bit out of the way with respect to the bi-partial paradigm. This applies, naturally, first of all to the intuitive obviousness of the bi-partial construct. Yet, it turns out that this case not only is fully in formal and procedural agreement with the bi-partial paradigm, but also provides quite a broad perspective on the problem analysed, which can, in general, be perceived and treated in a vast multitude of ways. The bi-partial paradigm is also of assistance in the analysis of this problem.

References

Bury, H., & Wagner, D. (2002). Kemeny's median algorithm. *Application for determining group judgement. CSM'2002* Laxenburg: IIASA.

Kemeny, J., & Snell, L. (1960). *Mathematical models in the social sciences*. Boston: Ginn.

Marcotorchino, F., & Michaud, P. (1979). *Optimisation en Analyse Ordinale des Données*. Paris: Masson.

Marcotorchino, F., & Michaud, P. (1982). Aggrégation de similarités en classification automatique. *Revue de Stat. Appl., 30,* 2.

Owsiński, J. W. (2011). The bi-partial approach in clustering and ordering: the model and the algorithms. *Statistica & Applicazioni* (pp. 43–59) (Special Issue).

Owsiński, J. W. (2012). Clustering and ordering via the bi-partial approach: the rationale, the model and some algorithmic considerations. In J. Pociecha & Reinhold Decker (Eds.), *Data analysis methods and its applications* (pp. 109–124). Warszawa: Wydawnictwo C.H. Beck.

Chapter 8
Final Remarks

This book is devoted to the possibly wholesome, but also possibly simple, presentation of the bi-partial approach in data analysis. The approach started from the attempt at a truly faithful rendition of the original problem of cluster analysis ("partition into subsets, inside which objects are as close to each other as possible, while those in different subsets are possibly distant"). The resulting general formulation of the objective function, expressed in two "dual" forms (1. to minimise the distances inside clusters + proximities among clusters, and 2. to maximise proximities inside clusters + distances among clusters) may make the impression of being banal, not promising any additional insight or technical capacity, regarding the solution to the clustering problem.

Yet, it turns out that starting from these very simple general precepts it is possible to:

a. *formulate a broad class of concrete objective functions*, applicable to various kinds of approaches and perceptions inside the domain of cluster analysis,
b. formulate analogous *functions and principles of analysis for a wide selection of problems in data analysis* outside of clustering,
c. show that these formulations, whether from cluster analysis or from other domains of data analysis, *imply non-trivial solutions*, not only *regarding the internal content of individual clusters* (or other constructs, obtained from the data), but also, very importantly, *their number*,
d. gain the powerful means for insightful modelling of the clustering problems, with distinction of different *levels of perception* (variables, objects, clusters, partition), with pronounced role of the *scale parameters*, appearing explicitly at the level of objects (transformation distance $\leftrightarrow$ proximity) and entire partition (the parameter of the objective function),
e. design an *effective algorithmic scheme*, leading to suboptimal solution, and that both in general form and for a wide class of concrete objective functions,

J. W. Owsiński, *Data Analysis in Bi-partial Perspective: Clustering and Beyond*, Studies in Computational Intelligence 818,
https://doi.org/10.1007/978-3-030-13389-4_8

f. demonstrate that this algorithmic scheme takes in many cases the form of the *classical agglomerative minimum distance procedure*, up to the possibility of associating the *Lance-Williams coefficients* to some of the respective algorithms.

Although the algorithms obtained feature the complexity not better (but also not worse) than the classical hierarchical merger procedures, like these procedures, they can be applied in the now quite popular hybrid approaches, if we wish to treat bigger data sets. It is also not excluded that for a narrower class of these algorithms some simplifications can be achieved, speeding up their functioning.

Concerning other potential research tasks to be undertaken in this domain we can mention the following ones:

(i) more detailed elaboration of the application of bi-partial approach to the various problems outside of the clustering domain (in this book only preliminary sketches are provided for some of them, requiring a much more profound treatment); likewise—potential application to yet other problems in data analysis, not mentioned here;
(ii) more rigorous scrutiny of the algorithmic conditions and the resulting potential characteritics of the algorithms; including also a better assessment of the range of potential reasonable algorithms, and
(iii) closer look at the relation between the bi-partial paradigm (the objective function and its versions, and the algorithmic precepts) and the classical hierarchical merger algorithms, based on the minimum distance principle, with the possibility of obtaining much more profound, also theoretical, results than those presented here.

It is true that the bi-partial approach does not apply to some of the important directions of work in cluster analysis, or, at least, the respective connections have not been established by the present author. It must be emphasised, though, that a part of these methods and algorithms are of the definitely local character, with no way to assign any (global) objective function to them within their proper paradigm. It can only be hoped that for all of these, for which an objective function can be formulated, some sort of bi-partial approach can be devised.

Leaving these and perhaps yet more open problems to the interested Readers the present author would like to stress again the simplicity and the vast capacity of the bi-partial approach, which ought to secure its proper place among the methodologies of data science.

Bibliography

This list contains the publications, concerning the development and applications of the bi-partial approach, which do not appear in the Reference lists, accompanying particular Chapters of the book.

Fedrizzi, M., Kacprzyk, J., Owsiński, J. W., & Zadrożny, Sł. (1994a). A DSS for consensus reaching using fuzzy linguistic majority and clustering of preferences. In Z. Nahorski & J. W. Owsiński, (Eds.), *Support systems for decision and negotiation processes*, special issue of *Annals of Operations Research* (Vol. 51, pp. 127–135).

Fedrizzi, M., Kacprzyk, J., Owsiński, J. W., & Zadrożny, Sł. (1994b). Group structure analysis for consensus reaching support. In R. Kulikowski, K. Szkatuła & J. Kacprzyk, (Eds.), *Proceedings of 9th Polish-Italian and 5th Polish-Finnish Symposium on Systems Analysis and Decision Support in Economics and Technology*, (pp. 80–90) Warszawa: Omnitech Press.

Kacprzyk, J., Owsiński, J.W., & Zadrożny Sł. (1993). Supporting consensus reaching using cluster analysis of the group's fuzzy preference structure. In *Proceedings of the First European Congress on Fuzzy and Intelligent Technologies*, (pp. 949–955) Aachen, Germany, 7–10 Sept 1993.

Kisiel-Łowczyc, B., Owsiński, J. W., & Zadrożny, S. (1999). Trade relation structures in Baltic Europe. *Argumenta Oeconomica, 2*(8), p. 30.

Kisiel-Łowczyc, A. B., Owsiński, J. W., & Zadrożny, Sł. (2000) Geographical structures of international trade in the Baltic Rim. In K. Jajuga & M. Walesiak, (Eds.) *Taksonomia 7. Klasyfikacja i analiza danych. Teoria i zastosowania*, Wyd. AE Wrocław, Materiały Konferencyjne PN 874, Wrocław, pp. 11–26.

Kulesza, B., Owsiński, J. W., & Sł, Zadrożny. (1989). Granulocyte luminescence curves: is anything there? *Applied Stochastic Models and Data Analysis, 5,* 329–339.

Owsiński, J. W. (1980). *Regionalization revisited: an explicit optimization approach. CP-80-26* (p. 24). Laxenburg: IIASA.

Owsiński, J. W. (1984). On aggregation in large scale models through a global clustering technique. In A. Straszak (Ed.), *Large scale systems: theory and applications* (pp. 207–212). Oxford: Pergamon Press for IFAC.

Owsiński, J. W. (1986). Optimization in clustering: an approach and other approaches. *Control and Cybernetics, 18*(2), 107–114.

Owsiński, J. W. (1989). On global optimality in cluster-wise regression. *Control and Cybernetics, 18*(1), 53–67.

J. W. Owsiński, *Data Analysis in Bi-partial Perspective: Clustering and Beyond*, Studies in Computational Intelligence 818,
https://doi.org/10.1007/978-3-030-13389-4

Owsiński, J. W. (1990). A simple software system for eliciting structured sets of notions from a group of experts (methods and experiences). In M. Schader & W. Gaul, (Eds.), *Knowledge, data and computer-assisted decisions*, (pp. 369–378). NATO ASI F61. Berlin-Heidelberg: Springer.

Owsiński, J. W. (1994a). Clustering and aggregation of fuzzy preference data: agreement vs. information. In E. Diday et al., (Eds.) *New approaches in classification and data analysis*, (pp. 478–481). Heidelberg: Springer.

Owsiński, J. W. (1994b). Aggregation and clustering of preferences: opinion vs. action. In A. Bachem, U. Derigs & M. Jűnger, (Eds.) *Operations Research '93*, (pp. 377–384). Heidelberg: Physica-Springer.

Owsiński, J. W. (1994c). Preferences, agreement consensus—measuring, aggregation and control. In: Z. Nahorski & J. W. Owsiński (Eds.), *Support for decision and negotiation processes*, special issue of *Annals of Operations Research*, (Vol. 51, pp. 217–240).

Owsiński, J. W. (1995a). What is out there to classify? In *Computer Data Analysis and Modelling*. Proceedings of the International Conference, Minsk, Sept. 4–8, 1995. Ministry of Education and Science of Republic of Belarus', Belarusian State University, Minsk, 106–115.

Owsiński, J. W. (1995b). Clustering—modelling, capacities, limits, applications. *Control and Cybernetics, 24*(4), 391–397.

Owsiński, J. W. (1996). Clustering, distances and knowledge from data. In E. Diday, Y. Lechevallier, & O. Opitz (Eds.), *Ordinal and symbolic data analysis* (pp. 277–287). Berlin-Heidelberg: Springer.

Owsiński, J. W. (1999). System of analysis of economic data on the micro level SAKIDOG. Outline and examples of results. *Taksonomia 6. Prace Naukowe AE we Wrocławiu, issue, 817,* 213–226.

Owsiński, J. W. (2001). Cluster-wise model identification. In S. Aivazian, Y. Kharin, H. Rieder, (Eds.), *Computer data analysis and modeling. robustness and computer intensive methods. Proceedings of the* 6th *International Conference (Sept. 10–14, 2001, Minsk)*, (Vol. 2, pp. 157–162). Belarusian State University, National Research Center for Applied Problems of Mathematics and Informatics, Belarusian Republican Foundation for Fundamental Research, Belarusian Statistical Association, Minsk.

Owsiński, J. W. (2002). Cluster-wise modelling: issues, proposals, methods. In K. Jajuga & M. Walesiak, (Eds.), *Taksonomia 9. Klasyfikacja i analiza danych—teoria i zastosowania, Prace Naukowe AE we Wrocławiu*, no. 942. Wyd. AE we Wrocławiu, Wrocław, 400–410.

Owsiński, J. W. (2003). Opinion structure and agreement conditions in group decision: the cluster analysis perspective. In J. Kacprzyk & D. Wagner (Eds.), *Group decisions and voting* (pp. 195–204). Warszawa: Akademicka Oficyna Wydawnicza EXIT.

Owsiński, J. W. (2004a). K-histograms: questions and perspectives. In S. Aivazian, P. Filzmoser & Yu. Kharin, (Eds.), *Proceedings of the* 7th *International Conference "Computer Data Analysis and Modeling. Robustness and Computer Intensive Methods", Minsk, September 6-10, 2004*, (Vol. 1, pp. 178–182). Academy of Administration at the President of the Republic of Belarus, Minsk.

Owsiński, J. W. (2004b). Distances, effective clustering and k-histograms: a natural way to produce linguistically meaningful clusters? In: P. Grzegorzewski, M. Krawczak & Sł. Zadrożny, (Eds.), *Soft computing. tools, techniques and applications* (pp. 219–227). Warszawa: Oficyna Wydawnicza EXIT.

Owsiński, J. W. (2005). Trade in hi-tech: a factor or effect in economic development of Baltic Europe. In H. Lindskog (Ed.), *Information technology in business. ITIB 2005* (pp. 135–148). 14–17 June, St. Petersburg, Russia. Linköping School of Management, Linköping universitet; Department of Informatics, St Petersburg State University of Economics and Finance, Linköping.

Owsiński, J. W. (2006a). Outlier detection: notions, problems, and methodological proposals. In K. Jajuga & M. Walesiak, (Eds.) *Taksonomia 13. Klasyfikacja i analiza danych - teoria i zastosowania, Prace Naukowe AE we Wrocławiu*, (pp. 45–55). no. 1126, Wyd. AE im. Oskara Langego we Wrocławiu, Wrocław.

Owsiński, J. W. (2006b). The ideal structures in group opinion analysis. In J. Kacprzyk & R. Budziński (Eds.), *Badania Operacyjne i Systemowe 2006* (pp. 207–217). Akademicka Oficyna Wydawnicza EXIT, Warszawa: Metody i techniki.

Owsiński, J. W. (2008). Ranking and ordering: some practical issues with a bearing on methodological and technical requirements. In J. W. Owsiński & R. Brüggemann (Eds.), *Multicriteria ordering and ranking: partial orders, ambiguities and applied issues* (pp. 253–269). Polish Academy of Sciences, Warszawa: Systems Research Institute. ISBN 83-894-7521-9.

Owsiński, J. W. (2013) On a bi-partial version of k-means. In B. Lausen, S. Krolak-Schwerdt, & M. Böhmer, (Eds.) *European Conference on Data Analysis 2013. Book of Abstracts*. University of Luxembourg, 133, ISBN 978-2-87971-105-8.

Owsiński J. W. (2010a) Asymmetric distances: potential output structures and procedures. *Studia i Materiały Polskiego Stowarzyszenia Zarządzania Wiedzą*, 31. Issue entitled "Dane – analiza – modelowanie – optymalizacja – podejmowanie decyzji", 317–325.

Owsiński J.W. (2010b) Asymmetric distances and fuzzy grouping. In: K. T. Atanassov et al., (Eds.) *Developments in Fuzzy Sets, Intuitionistic Fuzzy Sets, Generalized Nets and Related Topics*. Vol. II: Applications. SRI PAS/IBS PAN, Warsaw, 207–215.

Owsiński J.W., Kisiel-Łowczyc, A. B., & Kałuszko, A. (2001). Trade patterns, concentration, and dynamics: a preliminary analysis. In B. Johansson, D. Nilsson, (Eds.), *Trade and Transport Flows in the Baltic Sea Region. Proceedings, Workshop in Copenhagen, September 2000*. (pp. 32–54). Jönkoping International Business School, Working Paper Series, No. 2001–6.

Owsiński, J. W., & Pielak, A. M. (2008). Qualitative assessment of the websites of local authorities in Poland with hierarchical k-histograms. In K. Atanassov et al., (Eds.), *Developments in fuzzy sets, intuitionistic fuzzy sets, generalized nets and related topics. Applications*. (Vol. II, pp. 185–192). Warsaw: Academic Publishing House EXIT.

Owsiński, J. W., & Sł, Zadrożny. (1986). Clustering for ordinal data: a linear programming formulation. *Control & Cybernetics, 15*(2), 183–193.

Owsiński, J. W. Zadrożny, Sł. (1988a). Preference aggregation and ordinal grouping: the use of explicit objective functions. *Wissenschaftliche Zeitschrift*, Technische Hochschule Leipzig, 12, no. 3 part 2, pp. 177–185.

Owsiński, J. W., & Zadrożny, Sł. (1988b). A flexible system of precedence coefficient aggregation and consensus measurement. In A. Sydow, S. G. Tzafestas & R. Vichnevetsky, (Eds.), *Systems analysis and simulation 1988*, , Vol. 2: *Applications. Mathematical Research*, 47. (pp. 364–3704). Berlin: Akademie Verlag.

Owsiński, J. W., & Sł, Zadrożny. (1989). A decision support system for analysing and aggregating fuzzy orderings. In R. Kulikowski (Ed.), *Methodology and applications of decision support systems* (pp. 184–200). Warszawa: IBS PAN.

Owsiński, J. W., & Sł, Zadrożny. (1990). The problem of clusterwise aggregation of preferences. In R. Kulikowski & J. Stefański (Eds.), *Decision making models for management and manufacturing* (pp. 91–101). Warszawa: Omnitech Press.

Owsiński, J. W., & Sł, Zadrożny. (1991). Ecological site classification : an application of clustering. *Applied Stochastic Models and Data Analysis, 7*, 273–279.

Owsiński, J. W., & Sł, Zadrożny. (1992). Cluster-wise aggregation of relations: the case of paired comparisons of cognac ads. *Applied Stochastic Models and Data Analysis, 8*(2), 121–128.

Owsiński, J. W., & Zadrożny, Sł. (1995). Subjectivity vs. objectivity in cluster analysis: does it really matter, and, if so, why? *Statistics in Transition*, 2, 219–397.

Owsiński, J. W., & Zadrożny, Sł. (1997). Declarations and reality: clustering applied to voting of MPs in Polish parliament. In *Klasyfikacja i analiza danych. Teoria i zastosowania*. Sekcja Klasyfikacji i Analizy Danych PTS, *Taksonomia*, issue 4, pp. 166–179.

Owsiński, J. W., & Zadrożny Sł. (1998). Declarations vs. reality: what MPs say, how they are perceived and what they in fact do? In R. Kulikowski, Z. Nahorski & J.W. Owsiński, (Eds.), *Transition to Advanced market institutions and economies* (pp. 320–323). Conference materials, IBS PAN, Warszawa.

Owsiński, J. W., & Zadrożny Sł. (2000a). Structuring the set of MPs in Polish parliament: a simple clustering exercise. In Z. Nahorski, J. W. Owsiński & T. Szapiro, (Eds.), *Transition to advanced market institutions and economies*, special issue of *Annals of operations research*, 97, Baltzer, pp. 15–29.

Owsiński, J. W., & Zadrożny, S. (2000b). Suicide rates and their patterns vs. measurement of quality of life. In W. Ostasiewicz, (Ed.), *Aspects of quality of life* (pp. 263–285). Wrocław: Wrocław University of Economics Publishing House.

Owsiński, J. W., & Sł, Zadrożny. (2004). Trade in hi-tech products around the Baltic Rim: a business as usual or a progressive phenomenon? In J. W. Owsiński (Ed.), *MODEST 2004: Integration, trade, innovation & finance: from continental to local perspectives* (pp. 85–96). Warszawa: PTBOiS.

Owsiński, J. W., & Sł, Zadrożny. (2005). Tracing globalisation through gravity-based trade models: problems and suggestions. In J. Kacprzyk, Z. Nahorski, & D. Wagner (Eds.), *Zastosowania Badań Systemowych w Nauce, Technice i Ekonomii* (pp. 197–211). Warszawa: Akademicka Oficyna Wydawnicza EXIT.

Index

J. W. Owsiński, *Data Analysis in Bi-partial Perspective: Clustering and Beyond*, Studies in Computational Intelligence 818,
https://doi.org/10.1007/978-3-030-13389-4

Printed in the United States
By Bookmasters